⌐JUN 2 0 2007

EAST MEADOW PUBLIC LIBRARY

3 1299 00765 0626

A PLUME

THE LAST MAN WHO KNEW EVERYTHING

ANDREW ROBINSON is a King's Scholar of Eton College and holds degrees from Oxford University (in science) and the School of Oriental and African Studies, London. He is the author of more than a dozen books, including four biographies: *Einstein: A Hundred Years of Relativity; The Man Who Deciphered Linear B: The Story of Michael Ventris; Satyajit Ray: The Inner Eye;* and *Rabindranath Tagore: The Myriad-Minded Man* (written with Krishna Dutta). Since 1994, he has been the literary editor of *The Times Higher Education Supplement* in London.

"Anyone interested in what it mea⁀⁀ ⁀⁀ ⁀⁀⁀⁀⁀⁀⁀⁀⁀⁀⁀⁀⁀⁀⁀⁀⁀⁀ ⁀⁀⁀⁀⁀ ok."
⁀⁀cientist

"Thoroughly researched . . . Robi⁀⁀⁀⁀⁀⁀⁀⁀⁀⁀⁀⁀⁀⁀⁀⁀⁀⁀ o lived in a time when the fields of know⁀⁀⁀⁀⁀⁀⁀⁀⁀⁀⁀⁀⁀⁀⁀⁀⁀⁀⁀⁀⁀⁀ *Weekly*

"Young's polymathy gave him a u⁀⁀⁀⁀⁀ understanding of the world but also damned him to be forgotten by it. Until now, of course." —*Seed* magazine

"Thomas Young was not someone you'd want to go up against on *Jeopardy!*"
—*Kirkus Reviews*

"Reminds us how most of us tap only a small proportion of our full potential. It is also a cautionary tale on how society reacts to individuals who cannot be pigeonholed." —Arthur C. Clarke, author of *2001: A Space Odyssey*

"Egyptologist, epigrapher of genius, polymath, and life insurance expert, Thomas Young was one of the last true scientific generalists. . . . This skillfully wrought, long overdue biography is a masterly assessment of the man and his life." —Brian Fagan, archaeologist and author of *The Rape of the Nile*

"Thomas Young was both physicist and physician, practicing medicine throughout his life. . . . This is a fascinating and very readable account of the polymathic Dr. Young.
—Chris McManus, professor of psychology and medical education, University College London, and author of *Right Hand, Left Hand*

"It is the best biography I have read for many years." —Sir Patrick Moore, Fellow of the Royal Society, astronomer and writer

"It is wonderful to have such an elegant biography of this remarkable man."
—Philip Anderson, Nobel Laureate, Princeton University

JUN 2 0 2007

Portrait of Thomas Young in the 1820s.

The Last Man Who Knew Everything

Thomas Young, the Anonymous
Genius Who Proved Newton Wrong
and Deciphered the Rosetta Stone,
Among Other Surprising Feats

Andrew Robinson

A PLUME BOOK

PLUME
Published by Penguin Group
Penguin Group (USA) Inc., 375 Hudson Street, New York, New York 10014, U.S.A. • Penguin Group (Canada), 90 Eglinton Avenue East, Suite 700, Toronto, Ontario, Canada M4P 2Y3 (a division of Pearson Penguin Canada Inc.) • Penguin Books Ltd., 80 Strand, London WC2R 0RL, England • Penguin Ireland, 25 St. Stephen's Green, Dublin 2, Ireland (a division of Penguin Books Ltd.) • Penguin Group (Australia), 250 Camberwell Road, Camberwell, Victoria 3124, Australia (a division of Pearson Australia Group Pty. Ltd.) • Penguin Books India Pvt. Ltd., 11 Community Centre, Panchsheel Park, New Delhi – 110 017, India • Penguin Group (NZ), cnr Airborne and Rosedale Roads, Albany, Auckland 1310, New Zealand (a division of Pearson New Zealand Ltd.) • Penguin Books (South Africa) (Pty.) Ltd., 24 Sturdee Avenue, Rosebank, Johannesburg 2196, South Africa

Penguin Books Ltd., Registered Offices: 80 Strand, London WC2R 0RL, England

Published by Plume, a member of Penguin Group (USA) Inc. Previously published in a Pi Press edition.

First Plume Printing, January 2007
10   9   8   7   6   5   4   3   2   1

Copyright © Pearson Education, Inc., 2006
All rights reserved

The engraving opposite the title page is taken from the frontispiece of Young's last book, *Rudiments of an Egyptian Dictionary in the Ancient Enchorial Character* (1831).

 REGISTERED TRADEMARK—MARCA REGISTRADA

The Library of Congress has catalogued the Pi Press edition as follows:

Robinson, Andrew, 1957–
    The last man who knew everything : Thomas Young, the anonymous polymath who proved Newton wrong, explained how we see, cured the sick, and deciphered the Rosetta stone, among other feats of genius / by Andrew Robinson.
        p. cm.
    Includes bibliographical references and index.
    ISBN 0-13-134304-1 (hc.)
    ISBN 978-0-452-28805-8 (pbk.)
    1. Young, Thomas, 1773–1829. 2. Scientists—Great Britain—Biography. 3. Physicians—Great Britain—Biography. 4. Linguists—Great Britain—Biography. 5. Science—Great Britain—History—18th century. 6. Science—Great Britain—History—19th century. 7. Discoveries in science—Great Britain—History—18th century. 8. Discoveries in science—Great Britain—History—19th century. I. Title.
    Q143.Y7R63 2006
    509'.2—dc22   2005026912

Printed in the United States of America

Without limiting the rights under copyright reserved above, no part of this publication may be reproduced, stored in or introduced into a retrieval system, or transmitted, in any form, or by any means (electronic, mechanical, photocopying, recording, or otherwise), without the prior written permission of both the copyright owner and the above publisher of this book.

PUBLISHER'S NOTE
The scanning, uploading, and distribution of this book via the Internet or via any other means without the permission of the publisher is illegal and punishable by law. Please purchase only authorized electronic editions, and do not participate in or encourage electronic piracy of copyrighted materials. Your support of the author's rights is appreciated.

BOOKS ARE AVAILABLE AT QUANTITY DISCOUNTS WHEN USED TO PROMOTE PRODUCTS OR SERVICES. FOR INFORMATION PLEASE WRITE TO PREMIUM MARKETING DIVISION, PENGUIN GROUP (USA) INC., 375 HUDSON STREET, NEW YORK, NEW YORK 10014.

FOR DIPLI,
"CON AMORE"

Contents

Preface

Versatile people have always fascinated me as a biographer. Most recently, there was Albert Einstein, who, as everyone knows, fathered diverse new fields of science, but who also influenced some crucial areas of international politics. Before Einstein, Michael Ventris, a professional architect who in his spare time deciphered Linear B, the earliest European writing system, and became revered by archaeologists. And before Ventris, two prodigious Indians, the writer Rabindranath Tagore and the filmmaker Satyajit Ray, both of whom were intensely creative in areas outside literature and cinema.

But I must admit that Thomas Young (1773-1829), for sheer range of expertise, beats them all. Not only did he make pioneering contributions to physics (the wave theory of light) and engineering (the modulus of elasticity), to physiology (the mechanism of vision) and to Egyptology (the decipherment of the hieroglyphs), but he was also a distinguished physician, a major scholar of ancient Greek, a phenomenal linguist, and an authoritative writer on all manner of other subjects, from carpentry and music to life insurance and ocean tides. In an exhibition on Young arranged by London's Science Museum for his bicentenary in 1973, the organizers went so far as to state: "Young probably had a wider range of creative learning than any other Englishman in history. He made discoveries in nearly every field he studied".

This makes Young a tough subject for a biographer, and perhaps that is why there has not been a new biography of him for half a century. I have contemplated writing one for over a decade, since first encountering Young while researching a book, *The Story of Writing*, and I became further committed to the idea while writing another book, *Lost Languages*, on archaeological decipherment, a few years later. But having thought about the challenge, I decided it would be better to write an introduction to Young for a new audience, rather than attempting a full biography. To cover his work and life in detail and with authority is probably impossible for a single writer. This book therefore dwells only on the highlights of his polymathic career, though it aims to touch on every interesting and enduring aspect of Young.

I should like to thank the following for their help. Nicholas Wade, professor of visual psychology at Dundee, procured for me a four-volume set of the recent facsimile edition of Young's most famous work, *A Course of Lectures on Natural Philosophy and the Mechanical Arts*, for which he wrote an introduction. Christina Riggs, curator of Egyptology at the Manchester Museum, advised me on Horapollo's hieroglyphs. David Sprigings, consultant cardiologist at Northampton General Hospital, encouraged me to trace the post-mortem examination of Young to the library of St George's Hospital, London (where Young was a physician), and provided an expert opinion on the cause of his early death. Simon Young, son of the physiologist J. Z. Young, and great-great-great-grandnephew of Thomas Young, kindly gave me permission to reproduce his copy of the portrait of his ancestor painted by Sir Thomas Lawrence. Finally, I am grateful to my publisher, Stephen Morrow, at Pi Press, for getting excited by Young's versatility, too.

London, September 2005

Introduction

"Fortunate Newton, happy childhood of science! ... Nature to him was an open book, whose letters he could read without effort. ... Reflection, refraction, the formation of images by lenses, the mode of operation of the eye, the spectral decomposition and the recomposition of the different kinds of light, the invention of the reflecting telescope, the first foundations of color theory, the elementary theory of the rainbow pass by us in procession, and finally come his observations of the colors of thin films as the origin of the next great theoretical advance, which had to await, over a hundred years, the coming of Thomas Young."

Albert Einstein, Foreword to the fourth edition of
Isaac Newton's Opticks, 1931

Open any book on the science of light and vision, and you cannot miss the name of Thomas Young. At the very beginning of the nineteenth century, Young first demonstrated the interference of light by shining a beam of light through two narrow slits and observing the pattern the split beam created on a screen. "Young's fringes", as they became known, showed that light added to light could produce more light—or, most surprisingly, darkness. This pattern could be satisfactorily explained only on the basis that light was an 'undulation', that is, a wave—not a stream of 'corpuscles', that is, particles, as maintained by Isaac Newton in his deeply influential *Opticks*, published a century before Young. Over the next few decades, the undulatory theory of light totally supplanted Newton's corpuscular theory; and in the second half of the century, light was reconceived purely as an electromagnetic wave. Then, in 1905, Albert Einstein applied the quantum theory to radiation and discovered that light must be a stream of particles after all; shortly after, he became the first to make the revolutionary suggestion that both a wave theory *and* a particle theory of light could be correct. Today, a century after

Einstein's discovery, this has become the scientific orthodoxy, however puzzling: light, somehow, behaves as both a wave and a particle, depending on how you measure it. And Young's celebrated double-slits have become much more than a historically important experiment, since they can be used to demonstrate both wave and particle behavior. Repeated time and again by physicists with unimaginably more sophisticated and sensitive apparatus than Young's, the double-slit experiment encapsulates, said the physicist Richard Feynman, the "heart of quantum mechanics", its "only mystery".

But it is not only the physicists who claim Young as one of their own. He has an honored place in engineering, physiology and philology, too. Open any engineering textbook and you cannot fail to encounter "Young's modulus", a fundamental measure of elasticity derived from Hooke's law of stress and strain; Young's modulus is the ratio of the stress acting on a substance to the strain produced. Open any book on the eye, and Young will be there as the physiologist who first explained how the eye accommodates (in other words, focuses on objects at varying distances); who discovered the phenomenon of astigmatism; and, most important, who first proposed the three-color theory of how the retina responds to light, which was finally confirmed experimentally in 1959. Lastly, open any book on the languages and scripts of ancient Egypt, and Young is credited for some seminal detective work in deciphering the Rosetta Stone and the hieroglyphic script, which led to Jean-François Champollion's triumphant breakthrough in 1822. Even this great variety of achievements does not exhaust all that Young is remembered for, almost two centuries after his death in his mid-fifties in 1829. Far less important, though still noteworthy, are: "Young's rule" in medicine, a rule of thumb for deciding how to adjust an adult drug dosage for children; "Young's temperament" in music, a way of tuning keyboard instruments, such as harpsichords; and Young's principles of life insurance.

"Physicist, physician and Egyptologist" is how encyclopedias struggle to summarize Young. Physics and physiology were his forte, physic his profession, Egyptology his penchant. But his expertise extended well beyond these vast (even in his day) fields of knowledge. While not yet thirty years old, in 1802-03, as professor of natural philosophy at the newly founded

Royal Institution in London, Young gave a course of lectures covering virtually all of known science, which has never been surpassed in scope and boldness of insight, even by Michael Faraday, the brightest luminary of the Royal Institution; as a result, Young's lectures were reprinted as recently as 2002. No wonder he was elected a fellow of the Royal Society when he was barely 21, at the very outset of his medical training (and became its foreign secretary at 30 and, had he wished it, would most probably have been elected the society's president in 1827). If Nobel prizes had existed in the nineteenth century, Young would unquestionably have received one—perhaps even two Nobels, in physics for his work on the wave theory of light, and in physiology for his studies of the human eye and vision.

This was a man who, when pressed to contribute articles to a new edition of the *Encyclopaedia Britannica* in 1816, offered the following subjects: alphabet, annuities, attraction, capillary action, cohesion, color, dew, Egypt, eye, focus, friction, halo, hieroglyphic, hydraulics, motion, resistance, ship, sound, strength, tides, waves and "anything of a medical nature". Young was not boasting (he seems never to have boasted); for example, regarding annuities, he was a salaried "inspector of calculations" and physician for the Palladium Life Insurance Company, and regarding ships, he was an adviser to the Admiralty on methods of shipbuilding, secretary of the Board of Longitude, and superintendent of the vital *Nautical Almanac*. He did not bother to highlight to the encyclopedia's editor his polyglot knowledge of ancient and modern languages and of classical literature, especially ancient Greek (which helped him with the Rosetta Stone). In the event, he wrote authoritatively for the *Britannica* on many of the aforementioned subjects, plus articles on "Bathing", "Bridge", "Carpentry", "Double refraction", "Fluents" (integrals), "Herculaneum", "Languages", "Life preservers", "Road-making", "Steam engine" and "Weights and measures", as well as numerous biographies of eminent scientists and mathematicians and others such as his friend, the still-celebrated classical scholar Richard Porson. Young's three articles on "Egypt", "Languages" and "Tides" were far more than mere surveys of existing knowledge; they broke new ground, such as his coining of the term Indo-European to describe the family of languages first discovered in the 1780s, after he had compared the vocabulary and grammar of some four hundred languages.

A peculiarity of Young was that most of these contributions were anonymous; he feared that if he made public his multifarious scientific interests they would scare patients away from his medical practice. His instinct here was sound, even if he made too much of it, attempting to remain anonymous as a writer for most of his thirties and forties. In the class-conscious, comparatively unscientific, quack-infested medical world of his age—that of late Georgian London in the first two decades of the nineteenth century—Young's dazzling array of interests outside physic might well have given the impression of a doctor not wholly committed to his patients. Possibly the very breadth of his scientific knowledge may have sapped his confidence as a physician, given the primitive nature of medicine as a science. He certainly felt unable to advocate the vigorous medical treatments—copious blood letting, sweating, dieting and so on—favored by his confident medical contemporaries (and often by patients themselves), preferring to adopt more rational approaches to diseases. And while he was generally liked by men and charmed women, his reticence and modesty seem to have prevented him from exercising the 'bedside manner' expected of a consultant physician, a profession that demanded in his day a show of over-confidence to conceal chasms of ignorance. Whatever the reasons were, despite Young's being a respected physician at a famous London hospital, St George's, with prestigious medical lectures and publications to his credit, he never acquired the private practice that his scientific reputation should have warranted.

One of his letters from this time gives a fine idea of the variety of his interests and activities and some hints of the mind and personality behind them. It was written in his house in central London, just north of Oxford Street (today commemorated with a plaque), late in the evening of an unusually gloomy day in December 1820, when Young was 47 years old; and it was addressed to his oldest friend, the antiquary and politician Hudson Gurney, a fabulously wealthy man, who had confessed to Young in a letter that he was suffering from ennui and failure of resolution. ("Hudson's ambition was to write one good poem. Instead, he inadvertently became a multi-millionaire", notes a historian of the Gurney family.) Young replies to Gurney:

About this time last year I was giving myself a holiday of a few weeks, and I fell into a sort of fidgetty languor and fancied I was growing old; it went off very soon however, and I am convinced there is no remedy so effectual for this and other intellectual diseases as plenty of employment, without over-fatigue or anxiety. This autumn I have been in fact going on with the work which I was then almost frightened at having undertaken, "Elementary illustrations of the celestial mechanics of Laplace", and am already printing the first part of it— being only a translation with a commentary, it will do better without my name than with it. I am also writing over again my article on languages in the *Quarterly Review* with many additions for the next supplement of the *Encyclopaedia Britannica*—and a biographical memoir on Lagrange will be almost as long, requiring a list of 100 different papers on the most abstruse parts of the mathematics. I have then the business of the Board of Longitude to manage, and some of the Royal Society. The Arctic expedition is now settled; but we are fitting out our astronomer for the Cape of Good Hope with all his books and instruments, then there is a committee of elegant extracts to consider of the tonnage of ships, appointed by the Royal Society, the Admiralty, the Board of Trade, and the Treasury—which will not take long, but I shall have the onus—then there is my hospital—to speak modestly of my private patients—who are very discreet at this time of the year. By the way, such a day as this would make one glad to be anywhere rather than in London. I was forced to read by the fire and write in the dark at 1 o'clock: for I thought if I had candles I should scarcely have resolution to take my ride. Then I must not forget that I must very shortly fulfill my promise to do a little more to the hieroglyphics, and after one number more I shall be able to judge if the thing is worth continuing or not. ...

It is well for me that I have not to live over again; I doubt if I should make so good a use of my time as mere accident has compelled me to do. Many things I could certainly mend, and spare myself both time and trouble: but on the whole, if I had done *very differently* from what I have, I dare say I should have repented *more* than I now do of *anything*—and this is a tolerable retrospect of 40 years of one's life. ... I have learned more or less perfectly a tolerable variety of things in this world: but there are two things that I have never yet learned, and I suppose I never shall—to get up and to go to bed. It is past 12, and literally Monday morning as I have dated my letter, but I must write for an hour longer.

Just how truthful a picture of Young's life this was, is confirmed by a telling vignette of him from the same period written by one of his younger friends, Mary Somerville, the first woman scientist to win an international reputation in her own right (after whom Somerville College at Oxford is named). In her memoirs, she recalls how she and her husband and another couple (also known to Young) had been stargazing with a telescope until about two o'clock in the morning when they happened to notice a light in the window of Young's house in nearby Welbeck Street; clearly Dr Young was burning the midnight oil again. Mary's husband William, a former army doctor and a fellow of the Royal Society, rang the doorbell at No. 48, Dr Young appeared personally in his dressing gown, and they were invited inside to see a piece of Egyptian papyrus which he was then in the midst of translating. It appeared, said their brilliant friend, to be a Ptolemaic horoscope.

The only really accurate label for a man such as Young, apart from the overworked 'genius', would have to be 'polymath'. Or perhaps, 'Phenomenon' Young: the nickname given to him by Cambridge University students when he was resident at Emmanuel College in the late 1790s, apparently half in respect and half in derision.

Many of Young's distinguished contemporaries were willing to concede him the status of a unique polymath in his lifetime—though many others (including physicists, physicians and Egyptologists) were not, as we shall see. More significantly for us today, in the generation or two that followed him in the nineteenth century, Young's reputation climbed higher and higher among leading figures in the fields to which he had particularly contributed.

Lord Rayleigh, one of the giants of nineteenth-century physics (and the first British physicist to receive a Nobel prize), paid regular tribute to Young in his scientific papers, whilst also criticizing his writings for being too concise and thereby obscure. In 1899, lecturing at the Royal Institution on its centenary, Rayleigh did Young the signal honor of expounding some lesser-known aspects of his lectures given in 1802-03, which Young had published in his most famous book in 1807. According to the official record of the lecture, Rayleigh announced that: "Young occupied a very high place in the estimation of men of science—higher, indeed, now than at the time when he did his work. His *Lectures on Natural Philosophy* ... was a very remarkable book, which was not known as widely as it ought to be. Its expositions in some branches were unexcelled even now, and it contained several things which, so far as he knew, were not to be found elsewhere." Rayleigh concluded his lecture by noting that "possibly he had left the impression that Young knew everything. In fact, it was seldom that he was wrong; but just to show that he was, after all, human, a passage might be quoted from his book in which he declared there was no immediate connection between magnetism and electricity!"

Hermann Helmholtz, physiologist and physicist, another giant of his age, who stumbled across Young's forgotten three-color theory of vision and developed it in the 1850s into what is today known as the Young-Helmholtz theory, wrote famously:

> [Young] was one of the most acute men who ever lived, but had the misfortune to be too far in advance of his contemporaries. They looked on him with astonishment, but could not follow his bold speculations, and thus a mass of his important thoughts remained buried and forgotten in the *Transactions* of the Royal Society until a later generation by slow degrees arrived at the rediscovery of his discoveries, and came to appreciate the force of his arguments and the accuracy of his conclusions.

A third giant, the physicist James Clerk Maxwell, who extended the Young-Helmholtz theory, commented: "Thomas Young was the first who, starting from the well-known fact that there are three primary colors, sought for the explanation of this fact, not in the nature of light but in the constitution of man."

Young's status in Egyptian philology was more equivocal—partly as a result of his public controversy with the French scholar Champollion, which divided some people on nationalistic grounds—but not much less so than his prestige in physics and physiology. The Egyptologist François-Joseph Chabas, commenting on decipherment of the Rosetta Stone in the 1860s, wrote tersely of Young's chief contribution: "Cette idée fut, dans la réalité, le *Fiat Lux* de la science". In other words, Young's idea was the spark that created Egyptology as a science. A somewhat later Egyptologist, even more distinguished, Sir Alan Gardiner, called Young "a man of deep learning and wide interests, [who] was always ready to tackle any new puzzle"; Gardiner granted that Young would have completed the decipherment of the hieroglyphs he had started, had he but persisted instead of being sidetracked by his other interests.

In our own time, it is fair to say that Young's work is no longer at the forefront of science. Like any great scientific figure, parts of what he did have been incorporated into the foundations of science, and the rest is now the province of those interested in the history of science, whether they be working scientists or professional historians. Among both groups, it is not difficult to find examples of those who still marvel at Young as the scientists and scholars of the nineteenth century did.

For instance, an acclaimed book by a physicist, Arthur Zajonc, *Catching the Light: The Entwined History of Light and Mind*, remarks that Young was a "polymath of amazing reach" who "came to see himself as a modern Cassandra who spoke nothing but truth but whom no one could understand." An ophthalmologist, Gerald Fonda, laments in the *British Journal of Ophthalmology*: "Unfortunately for ophthalmology and for himself, Young was born 200 years too soon. Today he would probably be a giant in the research of physical, geometric, and physiological optics, enjoying success as a scientific physician, with a more appreciative and knowledgeable audience." Young's descendant, John Zachary Young, a zoologist and physiologist celebrated for his work on giant nerve fibers in squids, observes of his ancestor in his scientific autobiography that he was "the founder of all modern neurophysiology by his claims that the nerves carry information by their varying types." A professor of Egyptology at Cambridge University, John Ray, while paying all due respect to Champollion as the founder of his field, concludes a comparison of Young and Champollion with the striking comment: "The truth is that, in scientific discovery, the conceptual framework is the all-important first step. In Egyptology, that framework was the achievement of Young. ... Without Young's work, there might have been no study of ancient Egyptian."

However, as in his own day, Young also has a number of current detractors, each of them well-informed workers in their own fields.

For example, a physicist, David Park, in his highly praised *The Fire in the Eye: A Historical Essay on the Nature and Meaning of Light*, while discussing Young's work in generally laudatory terms, has this to say about Young himself:

> Like other members of the Royal Society, Thomas Young was a wealthy dilettante. Unlike most of them he possessed fabulous intelligence and a capacity for hard work, but he was a dilettante nonetheless. ... Though he wrote poetry in Latin and Greek he never learned to express himself clearly in English. ... Everything he wrote, and by report everything he said in his lectures, was vague and offhand.

A historian of science, Geoffrey Cantor, who has studied Young's note books for his 1802-03 Royal Institution lectures and written much on the lengthy disputes surrounding the wave theory of light, commented at the outset of his investigations in 1970: "Young studied medicine and dabbled in a wide range of subjects, rarely willing or able to concentrate on a single problem and work out its solution rigorously. It is clear that he was not trained as a natural philosopher. His knowledge of natural philosophy was gained almost exclusively through extensive reading." In 2004, Cantor concluded his frankly grudging entry on Young in the latest edition of the *Dictionary of National Biography*: "He was certainly highly intelligent but he appears to have lacked the discipline and insight necessary to pursue topics in great depth. He was most comfortable writing on subjects where he could organize the views of others in original ways."

Finally, a classicist, Maurice Pope, in his well-regarded *The Story of Decipherment: From Egyptian Hieroglyphs to Maya Script*, has little positive to say for Young as compared with Champollion:

> Young was a man with a grievance. After a brilliant youth ... he made original contributions to such diverse subjects as the theory of insurance, natural history, medicine, physics, and above all the history of technology, but never reached the first rank in any of them, except perhaps in optics in his work on the interference of light. Instead he rose to a position of considerable power in public life, becoming what would now be called a scientific and cultural administrator or adviser. Yet the rewards of this world did not satisfy him, and he

clearly hankered for something with a promise of immortality in it. ... [He did] useful enough work. It is a pity that Young spoilt it by laying claim to a glory that was not his. ... Even though everything that Champollion said [about Young's work] was both moderate and justified, time has inevitably made the details of the dispute seem trivial. It is a pity that there should be this slight tarnish, on one of the most important and original works of modern scholarship.

Plainly, there is a division of opinion among the experts. Those who appreciate Young admire his range, his intuition and his far-sightedness. Those who do not, depreciate these very same aspects of his life and work as dilettantism, sloppiness and opportunism. For the latter group, Young, far from being a polymath, stands convicted of some cardinal academic sins: lack of focus, lack of rigor and lack of originality. In a word, lack of discipline. Or should that be lack of *a* discipline? Two centuries after Young, in an age of narrow, and frequently narrow-minded, specialization in the academy and the professions unthinkable in his time, polymathy probably disturbs us still more than it did the Victorians. We are made uneasy—despite our cult of 'genius'—by those who effortlessly bridge several disciplines. It is only too natural to treat them as dilettantes or even to try to dismiss them as charlatans.

Up to a point, one sympathizes with the detractors. There can be little doubt that polymathy is exhausting, both for polymaths and for those who study them. Young himself died at the relatively early age of 55, which some at the time attributed to his incessant intellectual labors. The historian Alexander Murray put this point very perceptively in his introduction to an Oxford University symposium on the polymath Sir William Jones, 'Oriental' Jones, an Enlightenment figure of the generation before Young's, who died in 1794 at the early age of 47. Jones is principally known today— if he is known at all beyond some corners of the academy—for being the first person to identify clearly the similarities between Sanskrit, Greek, Latin, Gothic (Germanic), Celtic and Old Persian (the language family which Young then dubbed 'Indo-European'). Murray wrote on the bicentenary of Jones's death in 1994:

> History is unkind to polymaths. No biographer will readily tackle a
> subject whose range of skills far exceeds his own, while the rest of us,
> with or without biographies to read, have no mental 'slot' in which to

keep a polymath's memory fresh. So the polymath gets forgotten or, at best, squashed into a category we *can* recognize, in the way Goethe is remembered as a poet, despite his claim to have been a scientist, or Hume as a philosopher, for all the six dumpy volumes of his *History of England*. [Yet,] There are times when a mind of exceptional range, bestriding many conventional disciplines, makes a breakthrough in each because he knows the others, and all of them go on their way, afterwards, without necessarily recognizing what he did or how he did it. If history is not to be chronically misremembered, it follows that a constant effort must be made—as constant as the mechanism that pulls invisibly in the other direction—to recall those polymathic minds that have made these critical turns.

The Jones symposium required separate contributions from a Sanskritist and an Arabist, a theologian, a lawyer (Jones was a judge by profession) and an anthropologist, among others. No *one* of these specialists could really hope to judge the significance of all Jones's work. And the same is true in a recent move to revive the reputation of another polymath, Newton's contemporary Robert Hooke. A biography published for the tricentenary of Hooke's death in 2003, *London's Leonardo: The Life and Work of Robert Hooke*, was obliged to have four contributors from many disciplines, covering between them architecture, civil engineering, the history of science, natural history and social history. Again, this is probably inevitable, given Hooke's range. As Young himself, who was a deep admirer of Hooke, notes in the introduction to his 1807 Lectures: "A Boyle and a Hooke, who would otherwise have been deservedly the boast of their century, served but as obscure forerunners of Newton's glories." Young continues (without even mentioning Hooke's speculations on a wave theory of light):

> Hooke was as great in mechanical practice and ingenious contrivance, as Huygens was in more philosophical theory; he was the first that applied the balance spring to watches, and he improved the mode of employing pendulums in clocks; the quadrant, the telescope, and the microscope, were materially indebted to him; he had the earliest suspicions of the true nature of the cause that retains the planets in their orbits; and the multitude of his inventions is far too great to be enumerated in a brief history of the progress of science.

In some ways, Thomas Young is an even tougher biographical proposition than Hooke. First and foremost, his range was unquestionably greater than Hooke's, given that he worked in the humanities as well as in the sciences. Young surely has a better claim to be compared with Leonardo—as he has been—than Hooke has, especially given his deep knowledge of medicine. Secondly, Young's scientific work is considerably more mathematically sophisticated than Hooke's, though not by any means as sophisticated as that of contemporaries such as Pierre-Simon Laplace (for which Young would often be criticized by physicists as being unnecessarily imprecise). Thirdly, Young was an altogether more sociable, equable and appealing human being than the isolated and embittered Hooke, whose rancorous disputes with his contemporaries, especially the equally rancorous Newton, are legion. He was a lively, occasionally caustic letter writer, a fair conversationalist, a knowledgeable musician, a respectable dancer, a tolerable versifier, an accomplished horseman and gymnast, and, throughout his life, a participant in the leading society of London and, later, Paris, the intellectual capitals of his day. But while all this offers interesting material for the biographer, it often leaves Young's psychology in the shade—unlike Hooke's—because Young habitually cloaked his personal life, somewhat as he strove for anonymity in the authorship of his writings. We know, for example, very little about Young's relationship with his strict Quaker parents, and almost nothing about his wife Eliza except that their marriage was a happy one and she appreciated his work.

There have been two previous substantial biographies of Young: *Life of Thomas Young* by George Peacock, published by John Murray in 1855, which runs to almost 500 pages, and *Thomas Young: Natural Philosopher* by Alex Wood, published by Cambridge University Press in 1954, a somewhat shorter book. Peacock, who also edited Young's scientific papers, was a distinguished Cambridge University mathematician and professor of astronomy, a fellow of Trinity College (Newton's college), and also the dean of Ely (where he persuaded the chapter to undertake a complete restoration of the great cathedral). Wood, a lecturer in experimental physics, was also at Cambridge, though not as distinguished as Peacock; he died about two-thirds of the way through the writing of the book, which had to be completed by Frank Oldham, the author of an earlier brief portrait of Young published in 1933.

Both biographies have many merits, though neither could conceivably be described as an easy or lively read. Peacock was repeatedly requested to write the life by Mrs Young and was reluctant to agree, given his heavy professional commitments, illness and the daunting nature of the subject. He had access to Young's journals and private papers and the many frank letters Young wrote to Hudson Gurney—almost all of which have since disappeared, except for Young's manuscripts on his Egyptian research, which are in the British Library. Peacock's book is therefore invaluable for quoting at length from now-vanished original sources. On the other hand, Peacock is a prolix Victorian writer whose attempt to describe Young's scientific ideas entirely in words, without a single diagram (and, maddeningly, without any index), quickly becomes self-defeating. Wood, who also did much new research, writes more concisely and with greater clarity than Peacock, using excellent illustrations, yet his premature death deprived his book of coherence, and Wood lacks insight into (though not sympathy for) Young's personality. In both books, one cannot help but feel that the author was overwhelmed by Young's polymathy and abandoned the attempt to integrate the work and the life. Both Peacock and Wood struggle, and fail, to tell a story.

Which brings me to this book. My aim is simply to introduce Young's work and life to non-scientists who are unfamiliar with them. It would be futile and absurd to attempt comprehensiveness, for reasons which I trust are already obvious; instead I concentrate on the areas (suggested in the book's subtitle) where Young is considered by experts to have made definite discoveries and contributions, while trying to bring Young alive as a man. I also stress, unlike Peacock and Wood, Young's role as a physician, because this apparently influenced almost everything he did, even though he was not a conventionally successful medical practitioner.

I said earlier that Young's modern academic detractors deserve at least some sympathy, given Young's spectacular spectrum of interests. The title of this book, *The Last Man Who Knew Everything*, should hardly be taken literally. However, my prevailing sympathies will be obvious to readers.

Without overlooking its drawbacks, I find Young's polymathy awe-inspiring. It seems to me that Young was the opposite of the dilettante alleged by his detractors, if by that word they mean "a person who takes an interest in a subject merely as a pastime and without serious study, a dabbler" (as defined in the *Oxford English Dictionary*). But Young might well have allowed the word's Latin root as a fair description of his deepest motives as a scientist and scholar: *delectare*, "to charm, delight, amuse". On his deathbed, he was still working with a pencil, unable to manage a pen, on the final proofs of his path-breaking *Rudiments of an Egyptian Dictionary in the Ancient Enchorial Character*, which he did not quite live to finish. When a friend expostulated that this activity would fatigue him, "he replied that it was no fatigue, but a great amusement to him". If Thomas Young was a dilettante, then so, I submit, was Leonardo da Vinci.

Chapter 1

Child Prodigy

"Although I have readily fallen in with the idea of assisting you in your learning, yet [there] is in reality very little that a person who is seriously and industriously disposed to improve may not obtain from books with more advantage than from a living instructor ... Masters and mistresses are very necessary to compensate for want of inclination and exertion: but whoever would arrive at excellence must be self-taught."

Young, letter to his brother, 1798

Two or three years before his death, Thomas Young wrote a substantial autobiographical sketch in the third person, intended to be of use to someone writing an entry on "Young, Thomas" in a future edition of the *Encyclopaedia Britannica*. Possibly he had yielded to the idea at the request of his favorite sister-in-law, Emily, to whom he gave the manuscript. Immediately after his death, it was consulted by his friend Hudson Gurney in writing his brief memoir of Young, and again in the 1840s and 50s by his biographer George Peacock; then it disappeared. It was rediscovered only in the 1970s in the papers of Sir Francis Galton at University College London inside a black folder marked "Biograph: notes whence extracts were made for *Hereditary Genius*." Galton, a scientist best known for his work on eugenics, had apparently consulted Young's sketch in the 1860s while researching his leading work, *Hereditary Genius: An Inquiry into its Laws and Consequences*, a book stimulated by the publication of his first cousin Charles Darwin's *On the Origin of Species* in 1859. For some reason, the manuscript was never returned to the Young family.

Young was not a good candidate for a hereditarian like Galton, who made virtually no use of the sketch in his book, for Young had no offspring and no eminent close relatives. While it certainly cannot be said of his immediate forebears that they were "wholly without distinction and wholly without learning"—as has been said of Newton's family by his biographer Richard Westfall—they do resemble Einstein's merchant ancestors in combining considerable material prosperity with little obvious distinction. Young's father, Thomas Senior, was a mercer (cloth merchant) and banker from the village of Milverton, near Taunton, in the county of Somerset, in the south-west of England, while his mother Sarah was the daughter of a respectable merchant also from Somerset. The only notable figure in the family was her uncle, Dr Richard Brocklesby, a well-connected London physician, who would later have a decisive effect on his great-nephew Thomas's life.

Young's autobiographical sketch is virtually silent on his parents and siblings. He notes that he was born in Milverton on 13 June 1773, the eldest of ten children, and makes no further reference to his brothers and sisters. Nor is there any mention of his mother other than her name. Of Thomas Young Senior, the son notes: "His father followed the commercial fashion of the day, and became a manufacturer of money: he was for a time very successful in his speculations: but though a man of strict integrity, he was at last involved in the ruinous consequences of the general depression of the value of landed property so fatal to the country bankers."

An obvious reason for Young's lack of warmth is that his parents were Quakers who regarded their nonconformist religion, with its prohibitions on attending places of entertainment and on frivolity, and its particular observances—such as wearing plain black dress and broad hats, and using the same terms of address, thee and thou, to all, regardless of rank—as very serious matters indeed. The parents were, according to Gurney, himself a Quaker who would have known the Youngs personally, "of the strictest of a sect, whose fundamental principle it is, that the perception of what is right or wrong, to its minutest ramifications, is to be looked for in the immediate influence of a Supreme intelligence, and that therefore the individual is to act upon this, lead where it may, and compromise nothing." There is no place at all in Quakerism for the established authority of the church and monarch.

Young—like his friend Gurney—ceased to be a Quaker in his mid-twenties while his parents were still alive, married a non-Quaker, and regarded himself as a member of the Church of England in adult life. But according to Gurney's memoir:

> To the bent of these early impressions he was accustomed in afterlife to attribute, in some degree, the power he so eminently possessed of an imperturbable resolution to effect any object on which he was engaged, which he brought to bear on every thing he undertook, and by which he was enabled to work out his own education almost from infancy, with little comparative assistance or direction from others.

However, Young's autobiographical sketch, on which Gurney's memoir was based, does not actually say this. What Young wrote—which was suppressed by Gurney and Young's biographer Peacock—was in fact distinctly ambivalent about, and even dismissive of, his Quaker roots:

> His parents were rather below than above the middle station of life: but they were members of the society of Quakers; among whom education is not only very equally distributed, but the perfect community of rights and pretensions, and the complete contempt of public opinion not only obviate a great part of the depression of the lower orders, but have a natural tendency to produce, in a person who has any consciousness of his own power, a sentiment not very remote from conceit and presumption.

Very likely, Young regarded his father and mother as sharing in this general self-righteousness verging on bigotry, and as a direct consequence took care in his own life always to keep a close check on any incipient feelings of conceit about his remarkable achievements. Yet still it does seem plausible to attribute at least part of his attraction to science, his industriousness, and his self-reliance to his Quaker upbringing. There was a disproportionately large number of Quaker physicians and scientists in eighteenth- and early nineteenth-century Britain, such as the physicians John Fothergill, Thomas Dimsdale and John Coakley Lettsom, the chemist John Dalton and the meteorologist Luke Howard. One reason was probably that "despite the emphasis on discipline", each member of the Society of Friends was "encouraged to form his or her own views on any subject", as noted by the historians John Brooke and Geoffrey Cantor in their survey of Quaker (and ex-Quaker) fellows of the Royal Society. Young himself notes in his autobiographical sketch: "if it was allowable to dwell more on one part than another of holy writ, he was most disposed to be impressed with the importance of that part which conjoins [enjoins?] the votaries of true and undefiled religion, to teach themselves unassisted from the writ."

A second reason for Young's lack of warmth toward his parents must surely have been that he was sent away from Milverton very soon after his birth, and thereafter never lived with his parents for more than periods of a few months. While this could well have been necessary because of lack of space in a small village house with a growing family, as suggested by Alex Wood, it does seem a surprising attitude for parents to take, especially toward their first-born son.

Thomas went to live with his mother's father Robert Davis, the merchant, who lived in Minehead, some fifteen miles from Milverton. He writes warmly in his autobiographical sketch of this grandfather, who strongly encouraged his grandson's education. He was fond of classical literature, and one of his favorite sayings, which made a lifelong impression on Thomas, consisted of the famous lines of Alexander Pope:

A little learning is a dang'rous thing;

Drink deep, or taste not the Pierian spring ...

Pieria was the legendary home of the Muses on Mount Olympus.

Very soon, by the age of two, Thomas was reading fluently. Before he was four, at the village school and at home with his aunt Mary, he had read the Bible twice through. He also began memorizing poetry in both English and Latin, even though he could not yet understand Latin properly. He taught himself to remember Oliver Goldsmith's entire poem *The Deserted Village*, which had been published a few years earlier, and his grandfather noted in a quarto edition: "This poem was repeated by Thomas Young to me, with the exception of a word or two, before the age of five." (Young himself, characteristically accurate, notes that he was then six.) He also read Jonathan Swift's *Gulliver's Travels* and Daniel Defoe's *Robinson Crusoe*, but made no comment on them; they seem to have been among the few English novels Young felt worthy of his careful attention, whether as a child or as an adult.

Like many child prodigies, his memory was formidable. Another such prodigy, from the previous generation, was Richard Porson, the classical scholar whose biography Young wrote for the *Encyclopaedia Britannica* after Porson's death. There he remarked that "though a strong memory by no means constitutes talent, yet its possession is almost a necessary condition for the successful exertion of talent in general, and, indeed, it is very possible that the other faculties of the mind may be strengthened by the early cultivation of this one." But Young added a significant rider on the subject of memory, thinking of the fact that Porson, for all his great scholarship in the classics, left nothing to the world that was truly original:

> It seems to be by a wise and benevolent, though by no means an obvious arrangement of a Creative Providence, that a certain degree of oblivion becomes a most useful instrument in the advancement of human knowledge, enabling us readily to look back on the prominent features only of various objects and occurrences, and to class them and reason upon them, by the help of this involuntary kind of abstraction and generalisation, with incomparably greater facility than we could do, if we retained the whole detail of what had been once but slightly impressed on our minds.

Looking back on himself as a child, Young wrote disarmingly: "As far as the qualities of the mind and feelings are concerned, he may be said to have been born old, and to have died young." From a very early age, it was clear to him that he wanted to study many different serious subjects at the most advanced available level, and he consciously set himself the task of mastering them. This was remarkable in itself, but what is more remarkable is that he did not lose his drive with increasing age: he remained curious and determined right to the end of his life, long after other child prodigies have burned themselves out. "I like a deep and difficult investigation when I happen to have made it easy for myself if not to all others," Young told Gurney in his forties, because "[it] keeps one alive." Perhaps he was fortunate that his Quaker relations did not believe in 'parading' a child prodigy. At any rate, said Isaac Asimov, the writer and scientist, of Young: "He was the best kind of infant prodigy, the kind that matures into an adult prodigy."

Not surprisingly, conventional schooling did not stimulate him. Before he was six, he was sent daily to a dissenting clergyman, "who had neither talent nor temper to teach any thing well". He also attended a "miserable" boarding school near Bristol for almost a year and a half. He was supposed to learn arithmetic there, but found that he had got to the end of the textbook under his own steam before his master had reached the middle with the class. However when he was almost nine, in 1782, he transferred to a school at Compton in Dorsetshire run by a Mr Thompson, which suited him better, because the headmaster allowed the pupils some freedom in the way they spent their time.

Here Thomas read the commonest Greek and Latin classics—Virgil, Horace, Xenophon and Homer—and some elementary mathematics, as well as acquiring some knowledge of French and Italian, using books published in Paris borrowed from a schoolfellow. Nothing amazing in this, given the emphasis on the classics in that period. But then he branched out in his language study into Hebrew, by reading for amusement a few chapters of the Hebrew Bible. He was soon hooked, and after finally leaving the school in 1786, aged thirteen, he devoted himself at home to Hebrew and read through 30 chapters of the Book of Genesis without any assistance. Then, in answer to a discussion started over the dinner table, as to whether there were as marked differences among eastern as among European languages, he began to learn Arabic and Persian. A neighbor who heard of his fascination with Oriental languages, though a complete stranger to him, lent him grammars of Hebrew, Chaldee, Syriac and Samaritan, and the "Lord's Prayer" written in more than a hundred languages, which gave him extraordinary pleasure. He also read through most of Sir William Jones's Persian grammar published in 1771.

It must have been around this time that Thomas—quaintly dressed no doubt as a country boy in his Quaker costume—was said to have been taken to visit London by a relative and to have become engrossed in reading a valuable classic at a bookseller's stall. The skeptical bookseller said something like "There, my lad, if you could but translate to me a page of that (valuable as it is) it should be yours." His young customer promptly turned the text into flowing English. The bookseller, true to his word, though wincing at his sacrifice, handed over the book.

If it is beginning to sound as if science was neglected in Young's early years, this was not so. The Compton school usher, Josiah Jeffrey, "a very ingenious young man", lent Thomas the *Lectures on Natural Philosophy* of Benjamin Martin, and an elementary introduction to the Newtonian philosophy. The optical part of Martin's book got him started on making telescopes and microscopes, initially with the help of Jeffrey, who also had "an electrical machine", which the boy was allowed to use frequently, though disappointingly Young does not reveal what such a machine, in those pre-Voltaic days, was for. Jeffrey took Thomas's help, too, with the grinding and preparing of various kinds of colors, which were available for sale to the boys and others, and with bookbinding; and he taught him the first principles of drawing, with which he copied several specimens from the copperplate of a book titled *The Principles of Design*. When Jeffrey left the school, Thomas took over and began selling paper, copperplates, copybooks and colors to his schoolfellows; he earned five shillings, useful pocket money in 1786.

During school holidays back in Milverton with his parents or in Minehead with his grandfather, Thomas developed these mechanical and scientific interests. His father had acquired at auction Joseph Priestley's book on air, which prompted a delighted Thomas to make his first chemical experiments. His father's neighbor, a land surveyor called Kingdon, was happy for Thomas to come and read at his house a three-volume folio edition of a dictionary of arts and sciences, and also to let the boy learn to use some of his mathematical and philosophical instruments. At Minehead, he got to know a saddler called Atkins, who kept a meteorological journal using a barometer and thermometer during the whole of 1782, which was published by the Royal Society in 1784. Atkins lent the boy a quadrant, "which became the constant companion of my walks", as he used it to measure the height of nearby hills, probably of Exmoor. Another productive encounter, with a man called Birkbeck, made Thomas passionate to study botany. So that he could see plants in detail, he decided to make a microscope following the instructions in Martin's book. He procured a lathe for turning the requisite optical glass and other materials, which he managed to get hold of with the help of his grandfather and a cooperative clerk working for his father. "My zeal for botany during these operations

was replaced by my fondness for optics, and subsequently by that for turning." At the same time, Martin's book introduced him to the method of
fluxions, as the Newtonian calculus was known—but left him frustrated at
not understanding it. "I well recollect," he wrote, "that having seen a
demonstration in Martin which exhibited, though unnecessarily, some
fluxional symbols, I never felt satisfied until I had read, a year or two afterwards, a short introduction to the method of fluxions." (Young's glancing
comment, "though unnecessarily", is typical of him; faintly pedantic in
such an anecdotal context, but at the same time reminding us of how he
would always disapprove of using unnecessary or ostentatious mathematics to describe scientific concepts and physical reality.)

But before his early education now starts to sound too much like that
of an archetypal scientist—a practical-minded boyhood obsessed with
making things and experimenting on nature to the exclusion of human
relationships—Young utters a perhaps surprising, cautionary sentence in
his autobiographical sketch: "not that he was ever particularly fond of
repeating experiments, or even of attempting new ones; for he thought the
sacrifice of time generally great, and the success very uncertain". Young did
like to use his hands and make experiments in the time-honored Royal
Society tradition of Newton, Boyle and Hooke, but he liked even more to
use his mind, by reading all the authorities on a subject and coming to his
own conclusion, which might lead him to an experiment of his own. In
this respect, Young was more like Einstein with his famous 'thought' experiments than Newton in his Cambridge college laboratory or the eighteenth-century anatomist John Hunter, the 'father of modern surgery',
whose last lectures Young would attend as a medical student. "[Hunter's]
early distrust of the written word would make him forever skeptical of
classical teaching and the slavish repetition of ancient beliefs; he would
always prefer to believe the evidence of his eyes to the written words of
others," writes Hunter's biographer, Wendy Moore. Young, by contrast,
respected both kinds of evidence—though some of his later critics would
say that as a result of this lack of enthusiasm for experiments and his
respect for the literature, he did not conduct enough experiments and
lacked originality as a thinker.

The next five years of his life, from 1787 to 1792, before he became a
medical student, Young himself thought were "perhaps the most profitable

of his life, with regard to mental and moral cultivation". He was transplanted from rustic Somerset to a country house known as Youngsbury, near Ware in Hertfordshire, not far to the north of London, where he would spend two-thirds of each year, and the other third in a house at Red Lion Square in central London, for the duration of the winter months. He rarely went home to Somerset.

The abrupt shift was the result of Quaker family networking. Youngsbury belonged to one of the wealthy Quaker banking and brewing families, the Barclays, as did the house in London. Its owner, David Barclay, was looking for a companion to share the education at home of his twelve-year-old grandson, Hudson Gurney, when his niece, Priscilla Gurney, happened to stay with Thomas Young's aunt and strongly recommended to her uncle the precocious, thirteen-year-old Thomas. Thomas's parents were in favor of the move, and it turned out to be a lucky perfect match for his personality and talents. He now became, in effect, part of the Gurney family, forming a lifelong friendship with Hudson Gurney despite theirs being an attraction of opposites. Thirty years later, in a letter, Young remarked to his friend Hudson:

> It is singular how much you and I are contrasted in everything: you are generally out of humor with yourself, though you have great reason to be satisfied with others: I am abundantly disposed to give due weight to my own merits, but I feel nothing like an obligation to the world in general whom I cannot persuade to swallow my prescriptions with as much docility as they drink your beer.

What Gurney appears to have lacked—resolution—Young had in spades. Throughout his life, Young was keen on the idea that what one man had done, another man could also do; he had only a small belief in individual genius. According to one story told by Gurney, the first time his friend mounted on horseback at Youngsbury, he tried to follow the groom over a six-bar gate and was thrown heavily to the ground. But he got up without saying a word—Gurney never saw Young lose his temper at any time in his life—and made a second attempt, was again unseated, yet this time managed to stay on the horse. At the third attempt, he cleared the gate. In years to come, said Gurney, after taking lessons in horsemanship, Young would show "all sorts of feats of personal agility".

The tutor appointed for the boys by Barclay declined to take the job, so Thomas simply took over the classical education of Hudson, and they began reading together the great models of classical antiquity in both Greek and Latin. A classical tutor, John Hodgkin, now arrived to keep an eye on the pair, but there was relatively little need for him. The result was eminently satisfactory to both sides: Hodgkin was able to pursue his own classical studies, while giving Young a few hints on his Greek penmanship, which in due course resulted in a teaching book of ancient Greek texts by Hodgkin, *Calligraphia Graeca*, with some beautifully written examples in Young's hand. The close and informed appreciation of the Greek letters picked up in this practice would much later prove invaluable when copying and analyzing the Rosetta Stone and ancient Egyptian manuscripts.

As for mathematics and the sciences, the years at Youngsbury seem to have passed in extensive reading, rather than the practical experimentation of Young's first years. He studied botany and zoology and was particularly fond of entomology, but devoted much effort to reading mathematics, including Newton's *Principia* and *Opticks*, which he tackled in 1790, when he was 17. His biographer Peacock, a Cambridge tutor whose field was mathematics, was perplexed that a self-educated student could understand Newton's *Principia*, given the fact that it was a sealed book to most of Newton's mathematical contemporaries. But Peacock accepted Young's claim after reading his remarks in his private journal, which could only have been made by someone who understood Newton's propositions in detail. Nevertheless Peacock, who was severely critical of Young's general attitude to education—that private study was always superior to study in class with a teacher—was convinced that Young's method of learning mathematics was an inappropriate one, even for a mind as quick as Young's:

> A retentive memory and great clearness and precision of thought would appear to have superseded in his case the necessity of a more progressive training. In other respects the effects of this irregular intrusion into the inmost recesses of philosophy were such as might have been anticipated: he never felt the necessity nor appreciated the value of those formal processes of proof which other minds require.

Young kept a precise note of what he read each year. This is his list of books for 1790 in his own order with additional information in square brackets: "*Pentalogia Graeca* [by John Burton]; Reynolds's *Discourses*; *World* [a weekly published in 1753-56]; Lee's *Botany*; Bonnycastle's algebra; Sheridan on elocution; Haphaestion *Pauuri*; Linnaei [Linnaeus's] *Philosophia Botanica*; Homer's *Iliad* and *Odyssey*; Simpson's *Fluxions*; Corneille *chef d'oeuvres*; part of the *Monthly Review*; Virgil; Lettsom's *Fothergill*; Demosthenes of Mounteney; Custer and Leeds; Foster on accent and quantity; Blackstone's *Commentaries*; Hesiod; Aeschylus; Euripides; Sophocles; Newtoni *Principia*, Lycophron, Newtoni *Optica*; *History of France*, volume 3 (about 1789 and 1790) [by Nathaniel W. Wraxall]; part of Caesar, and of Cicero: Virgil, Horace; Juvenal, Persius; Terence; Sallust's *Catiline*; Martial, book 1; Eton Greek grammar; Greek Evangelists; *Cyropaedia* [by Xenophon]; part of Homer, of Euripides, of Sophocles, of Aeschylus, of Aristophanes; Rollin's ancient history; Gough's history of Quakers; Bonnycastle's astronomy; Euclid's six books; Boileau."

He was now on his way in right earnest to becoming a polymath. But Young himself was not especially impressed. In his autobiographical sketch, he said of his reading habits: "Though he wrote with rapidity, he read but slowly, [and] perhaps the whole list of the works that he studied, in the course of 50 years, does not amount to more than a thousand volumes: while it is said that William King the poet read no fewer than *seven* thousand in the course of a residence of *seven* years at Oxford." Of course, two centuries later, even literary scholars have hardly heard of this poet.

The only cloud over his intellectual idyll at Youngsbury appeared when Thomas was about fifteen. He glosses over it in his autobiographical sketch, but it must undoubtedly have been a cause of grave concern. He appeared to be developing a case of consumption—pulmonary tuberculosis—"a disease so frequent as to carry off prematurely about one fourth part of the inhabitants of Europe, and so fatal as often to deter the practitioner even from attempting a cure." This is from Young's grimly fascinating book, *A Practical and Historical Treatise on Consumptive Diseases*, published in 1815, which contains a number of observations on his personal symptoms in 1788-89. Perhaps the most interesting is the following:

[T]he dust of hard substances, constantly inhaled, seems to have an indisputable tendency to excite the disease: but the smoke of towns probably much less so than might naturally be imagined. In my own case, the symptoms originated in a very pure air, in a very healthy part of Hertfordshire, and subsided principally during a residence of some months in Red Lion Square, surrounded by closely built streets.

He also rejected on rational grounds, from personal experience, the supposition that once the tubercles had formed in the lungs, the consumption would become incurable:

I cannot help being persuaded that in my case there was an incipient formation of tubercles, the difficulty of breathing, and hectic symptoms, which I experienced, not being intelligible on any other supposition, since there was for a considerable time neither cough nor expectoration; and [the fact] that these tubercles must have disappeared at a subsequent period, was completely demonstrated by the restoration of the capacity of the chest to the extent of containing seven or eight quarts of air.

All that could be done for the youth was done. He was bled ("twice only"); a small blister was kept open on his chest for more than a year (at times "exceedingly painful"); a tonic was administered ("the Peruvian bark", that is cinchona containing quinine); and he was kept for two years on a diet of milk, buttermilk, eggs and vegetables with a very little weak broth ("little more than water in disguise"). His doctors were two well-known figures, both from Quaker backgrounds: Thomas Dimsdale, who had acquired a title, Baron Dimsdale, after inoculating Empress Catherine II of Russia against smallpox, and his great-uncle Richard Brocklesby, the physician of Samuel Johnson and Edmund Burke. With their assistance, and under the loving care of Mrs Barclay, Thomas made a complete recovery.

One of the few advantages of the disease was that it brought him to the attention of Brocklesby, who was somewhat beholden to Barclay. Judging from Brocklesby's letters, though, this must have been a mixed blessing for his great-nephew. Brocklesby comes across as a man of strong, and strongly held, opinions, about both medicine and the world, though affectionate in his own way and keenly aware of young "Tommy's" potential. Advising on the miserably dull diet, he writes in late 1789:

> Not that I am of opinion eating a little fish twice or thrice a week would hurt you, but you must make the trial cautiously and follow that which seems on experience not to be prejudicial. ... Recollect that the least slip (as who can be secure against error?) would in you, who seem in all things to set yourself above ordinary humanity, seem more monstrous or reprehensible than it might be in the generality of mankind. Your prudery about abstaining from the use of sugar on account of the Negro trade, in any one else would be altogether ridiculous, but as long as the whole of your mind keeps free from spiritual pride or too much presumption in your facility of acquiring language, which is no more than the dross of knowledge, you may be indulged in such whims, till your mind becomes enlightened with more reason.

Like many Quakers of his time, Young was an advocate of the abolition of slavery. On this occasion, he stood up to his elderly relative and well-wisher and continued his boycott of sugar and other products from the West Indian plantations—for in his autobiographical sketch, he notes proudly: "he was not fourteen when he took up the resolution of abstaining from the produce of the labor of slaves, and he adhered for seven years to this resolution, without once infringing it." Around the time he dropped it, in 1795, David Barclay of Youngsbury spent £3000 on liberating 30 slaves from a property in Jamaica that had fallen to him.

What really impressed old Brocklesby about his gifted great-nephew was the reaction to the young man's classical knowledge from the members of Brocklesby's circle of literary-minded friends in London. These included Edmund Burke, who had just published his influential *Reflections on the Revolution in France*; the statesman William Windham; Charles Burney, organist, composer and father of the novelist Fanny Burney; Sir Joshua Reynolds, the painter; and two physicians with strong interests in the classics, Dr Thomas Lawrence and Sir George Baker.

Young had sent his great-uncle a translation into Greek of some lines from Shakespeare's *Henry VIII*. Brocklesby wrote to Youngsbury with genuine enthusiasm:

I duly received a pleasing letter from you with a beautiful manuscript on vellum, a paraphrastic translation of Wolsey's farewell to Cromwell; better judges than I am, give it much praise for the spirit of Euripides, which they say it breathes ...

But Mr Burke has taken the Greek manuscript from me, and means to show it to divers learned men of his acquaintance for their philological criticism. I should be glad to have a copy of the same on vellum, as neatly written ... Mr Burke wishes you to try what you can make of Lear's horrid imprecations on his barbarous daughters ... If you can give the Greek the like compass of energetic expression as my favorite Shakspeare has done in his native tongue, Mr Burke will laud you and judge most favorably of your performance. He advises you to study Aristotle's *Logic*, his *Poetics*, and above all books, Cicero's moral and philosophic works. Your mind is not yet strained to any false principles, and he thinks you should be reared and cultivated in the best manner, so as to form your views, to emulate a Bacon or a Newton in the maturity and fullness of time; for he thinks it worth while for a comprehensive mind to be disregardful of any pecuniary emoluments of a profession, if you can but be satisfied with a small competence, and feel your mind prone to and satisfied with enlarged and useful speculations ...

Have a care, however, that my frankness towards you may not puff you up with vanity, which has been the rock that many others have split on, and I hope you will steer clear from ...

I had a fever since I last saw you, which has left exceeding weakness in my knees, so that I can hardly walk one hundred yards together, but I must learn to be satisfied in what is past. Pray God to have you under his immediate care, and that no imprudence of yours hereafter may frustrate the work that in you, with care, may be wrought.

Soon, the youthful prodigy was introduced to Brocklesby's intellectual circle in person. Thomas spent the last two months of 1791 staying with his great-uncle at his house off Park Lane, rather than at the Barclay house in Red Lion Square. In his journal for 12 December, Young noted that

Dr Lawrence, Sir George Baker, Richard Porson and another came to dinner; and that one of them read out Dr Johnson's Latin poem written on completion of his great dictionary. He recorded a conversation with Porson, who already admired Young's Greek penmanship, which began:

Young: Will *turba scholarum* do?

Porson: No; the five or six examples that may be brought are not sufficient to justify the making *à sch* short.

Young: What are we to make of *immensaque stagna*?

Porson: Most of the MSS have *immensa stagna*.

The rest of the conversation and other conversations show that Young could already hold his own in detailed discussion of Greek prosody. Meanwhile, the alcohol flowed convivially around the table—at least into Porson's glass if not into that of Young (as a Quaker he abstained), who remembered occasions when the subtlest nuances of Greek were discussed and dissected while Porson was "somewhat characteristically attempting to fill his glass out of an empty bottle".

Today's *Encyclopaedia Britannica* calls Porson: "British master of classical scholarship during the eighteenth century, the most brilliant of the English school that devoted itself to the task of freeing Greek texts from corruption introduced through the centuries." Young, in his *Britannica* entry on Porson, called him "one of the greatest men, and the very greatest critic, of his own or of any other age." And he explored Porson's achievement, thereby showing his own depth of classical scholarship, in considerable detail, while commenting that, "We find nothing in the nature of theory, or of the discovery of general laws, except some canons, which he has laid down, chiefly as having been used by the Greek tragedians in the construction of their verses." Young mentioned four such canons, of which the first two were: "when a tragic iambic ends with a trisyllable, or a cretic, this word must be preceded either by a short syllable, or by a monosyllable"; and that "an anapaest is only admissible in a tragic iambic, as constituting the first foot, except in some cases of proper names".

ΟΥΛΣΙΟΥ ΜΟΝΟΛΟΓΙΑ.

χαίροις ἂν ἤδη μακρὰ πᾶσ' ἐνδοξία·
χαίροιτε δυνάμεις, αἳ 'πισωρεύεσθέ μοι.
οὕτως ἔχει δὲ τἀνθρώπεια· σήμερον
ἀνὴρ τὰ χλωρὰ φύλλα τἀλπίδος φύει·
αὔριον ἀκμάζει, πορφυρέοις τ' ἐπ' ἄνθεσι
τιμῶν ὅσων περ ἔτυχε, πόλλ' ἀβρύνεται·
τρίταιον αὖτε ῥῖγος ἐμπίπτει βαρύ,
κἀπεὶ πεποιθὼς κάρτα γ' ἐλπίζει τάλας
καρπὸν μεγίστων ἐκπεπαίνεσθαι καλῶν,
ῥίζῃ πρὸς αὐτῇ δύσμορος ξηραίνεται,
κἄπειτα πίπτει δειλός, ὡς ἐγὼ τὰ νῦν.
ὁποῖα παῖδες νήπιοι παράφρονες,
ἐπὶ κύστεσιν νεῖν ἐν θέρει πειρώμενοι,
ὅπως τὰ πολλὰ γ' εὖκεκινδύνευκ' ἐγὼ
δόξης θαλάσσῃ, πρὸς βάθος μηδὲν σκοπῶν·
κόμπος δ' ἀραιὸς ὃν ἐπεφυσήκειν ἄταν
ἐσχισμένος λέλοιπέ μ' ἐν κλυδωνίῳ,
γέροντα, μόχθῳ καὶ χρόνῳ κεκμηκότα,
κἀνταῦθα λάβροις κύμασι βυθισθήσομαι.
ὦ λαμπρότητος καὶ τρυφῆς κενὴ σκιά!
ἀπεχθὲς ὄνομα! νῦν δὲ καρδίαν ἐμὴν
αὐταρχίας τυχοῦσαν εὖ γ' ἐπίσταμαι.
φεῦ δυστάλαιναν τοῦ τρισαθλίου τύχην
χάριτος τυράννων ὅστις ἐκκρεμάννυται!

Young's translation into Greek of the speech by Wolsey to Cromwell in Shakespeare's Henry VIII, *handwritten by Young on vellum, as shown in the biography of Young by Peacock.*

With such prowess in the classics, Young might have been expected to study Greek and Latin further at university, or to study law, which was the recommendation of Burke. But it seems that he was already drawn to physic—probably influenced by some of the leading physicians he had recently met—a profession that at the turn of the century was considered to require a classical training just as much as a scientific one. Moreover, Brocklesby, his great-uncle, who had no children, had made it clear that he would pay Thomas's way as a medical student and leave him part of his estate in his will, so he could set himself up as a London physician. To what extent Young was influenced by this tempting offer is unclear, but it appears to have caused some family tension, judging from a letter sent by Young's father from Milverton to David Barclay of Youngsbury in early 1791:

> [T]he plan for his studying physic is pretty generally approved of by his relations, and I hope not thought very unfavourably of by thyself ... I am apprehensive that the connection with his Uncle Brocklesby will add to his natural propensity to study and altho' the doctor is a man possessed of some valuable qualifications, yet I do not think him altogether fit to have the sole direction of young people therefore were he to make my son great offers I don't think it would be advisable to accept it, that is to the exclusion of myself and his other kind friends having the oversight of him, at same time I have no wish to offend him. ... If anything should occur I shall take it kind if thou wilt communicate thy sentiments to our Uncle Brocklesby as any remark from thee would be received much better than what *I* might say.

The following year, 1792, the decision was taken. Young spent his last summer of rural calm at Youngsbury. In the fall of that year, aged 19, he moved to London and took lodgings in Westminster not far from his medical lectures and his great-uncle's house (and not too close either, one may imagine Young as thinking). From now on, he became a citizen of the world's greatest metropolis.

Chapter 2

Fellow of the Royal Society

"It is well known that the eye, when not acted upon by any exertion of the mind, conveys a distinct impression of those objects only which are situated at a certain distance from itself; that this distance is different in different persons, and that the eye can, by volition of the mind, be accommodated to view other objects at a much lesser distance; but how this accommodation is effected, has long been a matter of dispute, and has not yet been satisfactorily explained."

Young, opening words of his first paper
read to the Royal Society, 1793

By 1800, London was "the best spot in Great Britain, and probably in the whole world where medicine may be taught as well as cultivated to most advantage", according to the well-known Bristol-based physician Thomas Beddoes. Even so, at the time when Young started his medical training in 1792, none of the city's six general hospitals had a medical school, and only St Bartholomew's Hospital offered regular medical lectures. However, there were many private lecture courses on offer, which must account for Beddoes's high praise of London medicine.

The most famous of these were the lectures at the purpose-built Hunterian school of anatomy in Great Windmill Street, Soho, just off Piccadilly Circus. Founded by William Hunter in the 1740s at a house in Covent Garden, where he was soon joined by his younger brother John, the school moved to the Soho precinct, then more salubrious, in the 1760s and in due course attracted ambitious would-be surgeons and physicians of all kinds. The Hunters believed that the only way for eighteenth-century surgery—the very phrase is chilling—to improve on its dismal record of fatal blundering was to study the human body minutely and learn the precise details of its internal organs and their interconnections. And the only way to do that was for students themselves to dissect corpses, not merely to watch a dissection done by another from a distance. In the Hunterian school, from its inception, a student was promised his own corpse—it was a key selling point of the lecture course. "In practice this meant that for every student who walked through [the] front door, another corpse needed to be heaved in through the back," writes John Hunter's biographer Wendy Moore. "And since a dead body rarely lasted much more than a week before decomposing beyond use, even in winter, in effect the school needed a steady stream of cadavers hustled through the back entrance on an almost nightly basis in order to keep the pupils coming."

There were two obvious sources of dead bodies: London's hangings and London's cemeteries. By royal authority, the Company of Barber-Surgeons had a right to six bodies annually from the public hangings at Tyburn Tree, the gibbet at the north-east corner of Hyde Park, where Oxford Street met Park Lane, near today's Marble Arch—just half a mile or so from the Norfolk Street house of Dr Richard Brocklesby. But to obtain control of the bodies hanging from the Tyburn scaffold, officials of the company were obliged to fight hand-to-hand battles with the relatives of the deceased convicts and with agents acting on behalf of anatomists like William Hunter, all the while surrounded by a crowd of Londoners unsympathetic to both the surgeons and the anatomists, who treated executions as a chance for a public holiday with all the fun of the festival. After the last public hanging at Tyburn in 1783, the only alternative for the anatomy schools—short of murder—was to make deals with corrupt undertakers and with body-snatchers, the so-called "resurrection men" who dug up fresh graves at dead of night and delivered the bodies to their clients well before dawn. Body-snatching grew fast during the century and had reached epidemic proportions by the time of Young's death in 1829, until it finally had to be controlled by the Anatomy Act of 1832. "Many a bereaved relative followed an empty coffin in a solemn funeral procession through Georgian London," writes Moore. "On several occasions when thefts were suspected, horrified relatives would frantically dig up grave after grave only to find every body gone."

The Hunters never explicitly mention body-snatching in their writings—in fact, there are virtually no eyewitness descriptions of it in print—and neither does Young, but he could hardly have been unaware of it, or of the primitive nature of much of the profession he was proposing to join. This was an age in which even leading surgeons like John Hunter thought nothing of operating with instruments still encrusted with the blood, pus and tissue of an earlier operation. Determined to improve surgery, Hunter was a controversial figure, both among fellow surgeons and physicians and among the wider public—so much so that when he died in late 1793, his colleagues at St George's Hospital voted not to send their condolences to his widow and even sponsored a posthumous biography damning him! Young, too, was fairly critical of Hunter, later writing in his book on consumptive diseases: "The works of Mr John Hunter exhibit many indications of a mind powerful and active, but not always subject to the laws of correct reasoning, and still less accustomed to be confined to clearness and precision of expression."

By the time that Young attended the Hunterian school, William Hunter had been dead for some years. John Hunter had ceased to lecture there in 1778, after a major public dispute with his brother that remained unresolved, but it appears that he cooperated with the school in later years, when it was run by his nephew Matthew Baillie and by William Cruikshank, both of them former pupils of his brother William. At any rate, during the autumn of 1793, Young attended lectures written by an ailing John Hunter and read by his brother-in-law and assistant Everard Home, during the course of which Hunter suddenly died on 16 October. At the end of his notes, Young recorded Home's valedictory comment: "We have gone through this course. It will never be repeated. It was only in hopes that Mr Hunter would have given practical lectures next winter. To keep the days open for him, I wished him to think himself pledged to go on. I mean to avail myself of his notes, and to give a practical course of operations next winter." A seemingly innocuous remark, but actually one with explosive implications. In what has become a notorious episode in medical history, Home, after Hunter's death, spent the rest of his prosperous medical career (leading to a baronetcy and the founder presidency of the Royal College of Surgeons) in plagiarizing Hunter's unpublished works for his own publications before burning the evidence in the 1820s.

As we shall see, Hunter and Home between them were a cause of major trouble for Young that came close to ruining his professional reputation. It is perhaps not surprising that Young mentions neither man in his autobiographical sketch, unlike Baillie, Cruikshank and his future medical lecturers at Edinburgh.

Around the same time, besides hearing two courses of lectures by these two anatomists at the Hunterian school, Young also entered himself as a pupil at St Bartholomew's Hospital, London's oldest hospital, founded in 1123, and located toward the original City of London in the east. Here he attended more lectures on various aspects of physic, including midwifery and botany, but also obtained his first bedside experience walking the wards with the physicians. While waiting for a lecturer to arrive, or perhaps bored by a lecture, he doodled on his notes. Being Young, his doodles were unusual: some were phrases in Latin and Greek, and others consisted of mathematical calculations and demonstrations. His non-medical interests persisted, almost unabated, while he was formally studying physic.

A dissection of an ox's eye he performed, presumably at the Hunterian school, gave him an idea for his first scientific paper. (His very first writing to be published, a short note on gum ladanum, appeared in 1791 in the *Monthly Review*, signed with his initials.) Young was intrigued by the way in which the eye focuses. Light rays from a near object need to be bent within the eye more than rays from a distant object, so as to be collected on the retina to form a sharp image of the object. "Accommodation is the process by which the eye can focus on objects at different distances, and it had been the dominant and unresolved topic in studies of vision for almost two centuries," writes the visual psychologist Nicholas Wade in his introduction to the 2002 reprint of Young's works.

Johannes Kepler had been the first to propose a theory of accommodation, in 1611, followed by René Descartes. In 1738, George Porterfield observed the vision of cataract patients who had had their crystalline lens *couched*, or removed, by an oculist. They could still see, but they could not accommodate; they could focus only with the help of convex glass lenses, and they required different glass lenses with different degrees of convexity in order to see objects at different distances distinctly. This showed that the crystalline lens, when focusing, must somehow change its orientation with respect to the retina. There seemed to be two possibilities for such change.

Either, the *position* of the crystalline lens must move back and forth within the eye along its horizontal axis like the lens of a camera (Kepler's theory), so that the image of a near object and the image of a distant object would be formed in the same plane on the retina, making each object appear in focus. Or, the crystalline lens must not move but stay in the same axial position in the eye, and instead its *shape* must change, becoming more or less convex depending on whether the object to be focused was near or far (Descartes's theory). A third possibility was that both processes might occur in accommodation: lens movement back and forth *and* change of lens curvature. Since Porterfield could find no evidence of muscle fibers in the crystalline lens—muscles being thought to be the only means of altering its curvature—he decided that the lens probably moved back and forth.

Having read all the relevant literature, noted particularly the effect of couching the crystalline lens, and then investigated the ox's eye carefully, Young cleaved to the Descartian view, not Porterfield's. He stated: "in closely examining with the naked eye, in a strong light, the crystalline from an ox, turned out of its capsule, I discovered a structure which appears to remove all the difficulties with which this branch of optics has long been obscured. On viewing it with a magnifier, this structure became more evident." After describing the crystalline lens's anatomy in some detail, he concluded: "Such an arrangement of fibers can be accounted for on no other supposition than that of muscularity." The new aspect of what Young proposed was that the lens itself was muscular, a fact that enabled it to change its curvature, in the way that earlier scientists had speculated.

He therefore advanced the following account of the accommodation of the eye:

> I conceive ... that when the will is exerted to view an object at a small distance, the influence of the mind is conveyed through the lenticular ganglion, formed from branches of the third and fifth pairs of nerves, by the filaments perforating the sclerotica, to the orbiculus ciliaris, which may be considered as an annular plexus of nerves and vessels; and thence by the ciliary processes to the muscle of the crystalline, which, by the contraction of its fibers, becomes more convex, and collects the diverging rays to a focus on the retina.

He was, we now know, absolutely right in his crucial conclusion. The human eye does accommodate by changing the curvature of its lens. But he was wrong in considering the lens itself to be muscular. In fact, the ciliary muscles, a set of radial muscles that surround the rubbery, jelly-like, non-muscular lens, are what alter the curvature. The function of the ciliary muscles was not known in Young's time, and so he attributed muscularity to the lens itself.

Young's paper, "Observations on vision", was read to the Royal Society on 30 May 1793 by Brocklesby, when his great-nephew was still only 19 years old, and was soon published in the society's *Philosophical Transactions*. Unfortunately for its young author, the immediate response was a claim by the great John Hunter to have already made the discovery and an application from him to the president of the Royal Society, Sir Joseph Banks, to give a major lecture on the subject during the following year. Though Hunter died before he could deliver this, the lecture was given in his place by his former assistant Home, even though Home himself did not endorse Hunter's 'muscular' lens.

At the same time, a rumor began to circulate that Hunter's idea had been discussed by the notably talkative Sir Charles Blagden at a dinner party given in November 1791 in the house of Sir Joshua Reynolds, with Young present along with Brocklesby, Thomas Lawrence, James Boswell and others. The implied plagiarism was so potentially damaging that Young immediately wrote to everyone who had been at the dinner and asked them if the subject of recent researches in vision had been mentioned. All were sure it had not been, except perhaps for the mischievous Blagden, who told the worried Young merely that he was "by no means so clear as to be sure that he told him Hunter's opinion." (We shall hear more of Blagden later.)

Much more reassuring was his election as a fellow of the Royal Society the following year, on 19 June 1794, a week after his 21st birthday. The proposal was made in March and described Young as "a gentleman conversant with various branches of literature and science, and author of a paper on vision". It was signed by, amongst others, the physicians Matthew Baillie, Sir George Baker, William Heberden Jr (physician to St George's Hospital and to George III), Everard Home and Brocklesby himself, by the co-founders of the Linnean Society, the naturalists William Shaw and

James E. Smith, and by Richard Farmer, master of Emmanuel College at Cambridge, where Young would later reside.

This was a real honor for one so young. But we should not be tempted to compare it with its modern equivalent. Before the middle of the nineteenth century, most of the fellows of the Royal Society, as for example Young's friend Hudson Gurney (elected an F.R.S. in 1818), did not possess any scientific credentials or publications, and even after this period, until the later decades of the nineteenth century, numerous of the fellows did not actively pursue scientific research. It is inconceivable today that even a young man as gifted as Young could be elected a fellow of the Royal Society on the evidence of one scientific publication.

Young's election scotched the plagiarism allegation but it did not quell his dispute with Hunter and his legacy. Home, who disagreed about the process of accommodation with both his former mentor Hunter and with Young, gave his lecture to the Royal Society in late 1794 in Hunter's place. He claimed to have shown, with the help of the leading optical instrument maker Jesse Ramsden, that 'aphakic' subjects— that is, patients with a couched eye or eyes —nonetheless had some power of accommodation— contrary to Porterfield's observations and of course Young's inferences from Porterfield. In the face of such evidence presented with such authority, Young decided that he had made a mistake and felt obliged publicly to withdraw his 'muscular' view of lens curvature. Only in 1800, after a series of brilliant experiments described in a later chapter, would he disprove Home's contention of 1794.

"I hope I am not thoughtless enough to be dazzled with empty titles which are often conferred on weak heads and on corrupted hearts," Young wrote to his Quaker mother in Somerset on being elected a fellow of the Royal Society. But while he seems largely sincere in this, given his later lack of interest in titles and his pursuit of anonymity as a writer, he was also obviously trying to reassure his mother that he was not moving away from his Quaker roots. No doubt his parents had noticed how much he had changed when he visited them in Milverton in May 1794 with Gurney while the two friends were en route to Cornwall to study its mines and mineralogy for a few weeks. In reality, Young knew he was being ineluctably drawn away from Quakerism. None of those with whom he was now associating in London were faithful Quakers. His father's earlier apprehension about "Uncle Brocklesby" had proved accurate.

In Bath, on his way to Milverton, Young stopped to visit his uncle's patient, the duke of Richmond, who was then taking the waters at the famous spa under the advice of Brocklesby. The duke was someone with whom a strict, pacifist Quaker should have little truck. He was a military man and politician, master-general of ordnance (military supplies) with a seat in the Cabinet. But the duke was keenly interested in scientific pursuits and well acquainted with surveying instruments through his interest in the great trigonometrical survey, which came under his department. He and Young hit it off, and the worldly Richmond wrote to Dr Brocklesby:

> I really never saw a young man more pleasing and engaging. He seems to have already acquired much knowledge in most branches, and to be studious of obtaining more: it comes out without affectation on all subjects he talks upon. He is very cheerful and easy without assuming anything; and even on the peculiarity of his dress and Quakerism he talked so reasonably, that one cannot wish him to alter himself in any one particular.

A further meeting with the duke at Goodwood in August 1794 led to an offer of a post as his assistant private secretary. This put Young in a quandary. He informed his mother that "a principal reason" for not accepting the proposal was his loyalty to Quakerism, but there is no mention of this reason in his 1820s autobiographical sketch. There he notes that he declined what he knew was an undoubted social advancement because "he had predilection for the more tranquil pursuits of science, which he thought more congenial to his talents and his habits". Nor is there any mention of Brocklesby's reaction, though Peacock notes that Burke and Windham, his great-uncle's friends, advised against accepting the offer. To have taken the position with the duke would presumably have put an end to his generous relative's future financial support. Whatever his true motives may have been, Young would remain on good terms with the duke, but would decide to continue on his way to the next stage of his medical training, first in Edinburgh and then in Germany.

Chapter 3

Itinerant Medical Student

"I expect many advantages from spending two years on the continent; not but that I believe almost all that can be known of physic might be learnt, if necessary in London. ... But besides that I by no means wish to confine the cultivation of my mind to what is absolutely necessary for a trading physician."

Young, letter to his mother from Edinburgh, 1795

At the turn of the eighteenth century, a university degree was not yet a *sine qua non* for a successful medical practitioner in Britain. John Hunter, famously, never studied at any university, though his brother William did, and neither did Hunter's assistant Everard Home, though his nephew Matthew Baillie did. But it was clear that a university qualification was becoming increasingly important as physicians attempted to distinguish themselves from quacks and create a profession out of medicine. For Young, though he had no strong interest in attending any educational institution—he writes almost nothing about his education at Edinburgh, Göttingen or Cambridge in his autobiographical sketch—a university degree was nevertheless a desirable qualification.

In choosing to study at Edinburgh University in 1794 and then at a continental university such as Leiden or Göttingen, Young was following in the footsteps of many other Quakers, including several fellows of the Royal Society, among them his great-uncle, Richard Brocklesby (who studied at Edinburgh and Leiden). As a physician in pursuit of the coveted 'M.D.' after his name, in one sense he knew he had no choice, since London had no university at this time and the ancient universities of

Oxford and Cambridge were closed to Quakers on religious grounds (and anyway did not yet possess medical schools). But he also knew that the path he had chosen made good professional sense. During the course of the eighteenth century, the medical training at Edinburgh University had become second only to that available in London's private academies and hospitals, if not superior in some departments, while a first-hand exposure to the continent, especially to the refinements of French, German and Italian scholarship and culture, Young knew was certain to give his scientific medical knowledge a socially acceptable finish of a kind that would be helpful in handling wealthy and aristocratic patients such as those of his great-uncle Dr Brocklesby. Perhaps, in the end, as important to him in his decision as all these reasons was simply his wish to broaden his knowledge even further than its already exceptional range.

In the days before there were railways, he had the option of traveling to Edinburgh either by horse-drawn coach or on horseback. Since he wanted to stop off on the way, and was already a highly experienced rider, he left London by horse, presumably carrying a number of books along with letters of introduction from his distinguished circle in London. In Derbyshire, the vicar of Buxton, who knew Dr Brocklesby, introduced him to Robert Bakewell, the pioneer in scientific methods of livestock breeding and husbandry. Young noted of his farm at Dishley:

> I felt his rams and sheep regularly as they were shorn, and went through all the forms of examination. What he has done is shown best by two sheep which have always lived together, one of his own improved breed, the other of his original breed from which all his stock was derived by selected mixtures without crossing with any other breed. He entirely neglects the wool, but has diminished the bone and increased the fat in a surprising degree. Some bones are cleaned as specimens, and some pieces of meat were hung up, four inches thick in fat.

At Derby, the interest was Erasmus Darwin, an eminent physician and biologist now known principally as the grandfather of Charles Darwin but in the 1790s celebrated for his *Zoonomia or the Laws of Organic Life*, the first volume of which had just appeared. Young was intrigued by it but already privately critical in his journal, and in his own book on consumptive diseases (published well after Darwin's death) wrote severely that,

"Much ingenuity, much practical knowledge, and much absurdity, are combined in the *Zoonomia* of Darwin. To follow his theories would be useless, but some of his hypothetical assertions require to be noticed, for their singularity and boldness." He very much enjoyed his visit to Darwin's house, however: "He gave me my choice of looking over three cabinets, of cameos, of minerals, and of plants; the two last I viewed very superficially, but spent some time with him in admiring a collection of impressions bought in Italy: he says that he borrowed much of the imagery of his poetry from the graceful expression and vigorous conception which they breathe." On departing, Young was given a letter of introduction from Darwin to a friend in Edinburgh that described his new young acquaintance most flatteringly: "He unites the scholar with the philosopher, and the cultivation of modern arts with the simplicity of ancient manners."

At Durham, he met a fellow student from his lectures in London, who was also on his way to attend the medical school at Edinburgh. "He has studied at Cambridge," Young noted, "and is well read in ancient and modern languages; he joins his knowledge with much modesty and agreeable dispositions." Together they called on a clergyman, who was also a Greek scholar, at the college next to the famous cathedral, but he was away: "I, however, left my name, with Hodgkin's plate of my Greek translation from *King Lear*."

He reached Edinburgh on 20 October, found lodgings in St James's Square, and was quickly immersed in studies and society. His reputation and his F.R.S. had preceded him, and the letters he carried completed the picture of unusual youthful brilliance.

The medical school at Edinburgh was established in 1726 with the appointment of Alexander Monro, who had done his training in Leiden, as professor of anatomy. He was succeeded by his son, also called Alexander, in 1758, and then by *his* son, a third Alexander, in 1798, who stayed until 1846. Thus the three Monros occupied the Edinburgh chair of anatomy for some 120 years. Other luminaries were Joseph Black, the chemist, William Cullen in theoretical medicine and John Gregory in practical medicine.

A staggering 17,000 medical students studied there in the first century of the school, notes the historian Roy Porter in his authoritative book, *The Greatest Benefit to Mankind: A Medical History of Humanity from Antiquity to the Present*. "The university had many attractions: it was cheap, there were no religious restrictions, and the lectures were in English"—rather than in Latin as at Oxford and Cambridge. "There was no obligation to graduate, students attended only the courses they desired, and paid for those alone. This demand-led regime kept the professors on their toes, and Edinburgh flourished because it offered the practical training students needed." What did this consist of? Porter describes it as:

> Not primarily an intensive bedside training. Though [Edinburgh] pioneered infirmary-based teaching, only around one third of the students signed up for clinical lectures. Nor did they get prolonged personal dissecting practice. There was always a shortage of corpses, and illegal acquisition of cadavers led to scandals which implicated anatomists and surgeons in body-snatching and finally murder. The strength of an Edinburgh education lay in imparting the elements of anatomy, surgery, chemistry, medical theory and practice. After three years, an Edinburgh man trained in medicine and surgery was ready to go out into the world to practice the new trade of 'general practitioner' or family doctor. Those falling sick in 1810 in Newcastle, Newfoundland or New South Wales would most probably have been seen by Edinburgh doctors.

Young, of course, had set his sights higher than on becoming a general practitioner. On the whole, though, he was appreciative of the standard of the many lectures he attended, singling out for mention those of Black, Gregory and Monro in his autobiographical sketch written much later. At the time he defended Edinburgh in a letter to Brocklesby from the criticisms of some of their circle in London: "with respect to the study of physic, it appears to me beyond comparison preferable to Oxford or Cambridge, and in other respects little inferior." One might ask how he could have known this, not having studied in either of those cities. He must have based his view—which was essentially correct, given the moribund state of medicine at Cambridge he would later encounter—mainly on talking to his fellow students at Edinburgh, five or six of whom were Oxford or Cambridge men.

When it came to the level of original thinking at Edinburgh, however, Young was much more critical. He thought none of the medical faculty to be of the first rank. About Monro, he was almost waspish. Monro claimed in his lectures to have been teaching the fibrous nature of the eye long before Hunter had announced it: "this was received with applause from his pupils, who always encourage his avarice of priority: in this case, though Monro deserves nothing, I was not displeased that Hunter's pretended originality was disallowed." Soon afterwards, Young called personally on Monro to inform him of his own paper published by the Royal Society and was told by Monro that he had recently been reading it and wished to study it further. Shortly after this visit, Home's opposing Royal Society lecture on the accommodation of the eye reached Young, and after studying it, he decided to tell Monro immediately that he was withdrawing his published opinion about the action of the crystalline lens. There was no response from Monro until near the close of the lecture course. Then he spoke of Young's paper "with as much respect as it deserved", and made some criticisms of it "which were partly worthy of attention, and partly groundless". But he ignored Young's change of mind, and passed over "in a very slight, and, I think, a very uncandid manner, the experiments stated in [Home's] lecture, insinuating, as is too common with him, that he had himself made observations of a similar nature."

It is difficult to know quite what to make of Young's complex reaction here. In due course, as mentioned in the previous chapter, he himself would prove Home's experiments to be wrong, though obviously he could not have known this when listening to Monro. So Monro was actually right to ignore Young's change of mind, but he did so for the wrong reasons, since he apparently agreed with Home's experiments. Presumably, therefore, Young was reacting with dislike to Monro's combination of slippery arguments and personal bravado. If so, it was a curious forewarning of a crucial future episode in his own scientific career, in which the same combination would occur in a devastating salvo from the *Edinburgh Review* against his evidence for the wave theory of light. (See Chapter 8, "'Natural Philosophy and the Mechanical Arts'".) For now, we should merely observe that Young could be extremely sensitive to what he perceived as unreasonable criticism—perhaps partly because he was an autodidact unfamiliar with the give-and-take of conventional education (a view strongly favored by his biographer Peacock), but more likely because most geniuses have a tendency to be very sensitive to criticism.

If his relationship with Edinburgh medicine was sometimes prickly, his contacts with other aspects of life in Edinburgh, then known as the Athens of the North, were a source of satisfaction and pleasure. His closest relationship was with Andrew Dalzel, the professor of Greek. As with John Hodgkin at Youngsbury and their *Calligraphia Graeca*, so with Dalzel at Edinburgh Young collaborated, and the result was a second edition of *Analecta Hellenika*, Dalzel's anthology of Greek poetry with selections from the epigrammists and learned notes by Young. Writing to a Professor Young of Glasgow, Dalzel remarked: "There is a namesake of yours here at present from England, a great Grecian, who has translated into Greek verse Lear's imprecation from Shakespeare." Young and Dalzel would correspond at intervals almost until Dalzel's death.

At the same time, in a city noted for its hospitality, Young took up theater-going, dancing and the flute. As usual, he was thorough in his approach. There is a story that some fellow students came into his room soon after his first dancing lesson on the minuet and found him tracing, with ruler and compass, the various crossings of the two dancers and how he thought improvements might be introduced. Soon after leaving Edinburgh, he wrote to one of these fellow students, Dr Bostock, with whom he was particularly friendly:

> I have seen Mrs Siddons in *Douglas*, *The Grecian Daughter*, *The Mourning Bride*, *The Provoked Husband*, *The Fatal Marriage*, *Macbeth*, and *Venice Preserved*. She was neither below, nor much above, my expectation. I can form an idea of something more perfect. … I know you are determined to discourage my dancing and singing, and I am determined to pay no regard whatever to what you say. You think I shall never be able to play the flute well, and I am pretty sure that I may if I choose; as to dancing, the die is cast.

The local Quakers, predictably, were not amused. One of them, "my friend Cruikshanks", took Young aside, and "after much preamble, told me he heard I had been at the play, and hoped I should be able to contradict it. I told him I had been several times, and thought it right to go, etc. etc., as civilly as I could." Here also the die was cast, and it was now only a matter of when, rather than whether, Young and Quakerism would part company.

While in Edinburgh, he read some relatively modern literature, in addition to his favored classical writers. With real satisfaction he worked his way through two romantic epics: Cervantes's *Don Quixote*, helped by a Spanish grammar and dictionary, and Ariosto's *Orlando Furioso* (a book his friend Gurney would much later translate from the Italian). And he read Samuel Johnson's *Rasselas* and his travel account of the Highlands in the 1770s, *A Journey to the Western Islands of Scotland*. This provoked a mixed reaction in Young's journal for 14 May 1795: "I began, and the next day finished Johnson's *Journey*. It exhibits some strength of mental powers, but with a mixture of pedantry, bigotry and prejudice. I have not extracted from it much information of what I may find in the Highlands, but the manners of the country are well depicted."

Three weeks later, with the Edinburgh medical lectures over, Young set off on his own Highlands journey, traveling alone by horse. Anyone familiar with Scottish weather will know that such a journey is not to be undertaken lightly. Young described the moment of departure from Edinburgh in his journal:

> I was mounted on a stout, well-made black horse, fourteen hands high, young and spirited, which I had purchased from my friend Cathcart: I had before me my oiled linens, the spencer with a separate camlet cover; under me a pair of saddle bags, well filled with three or four changes of linen, a waistcoat and breeches, materials for writing and for drawing, paper, pens, ink, pencils, and colors; packing-paper and twine for minerals; soap, brushes, and a razor; a small edition of Thomson's *Seasons*, a third flute in a bag, some music, principally Scotch, bound with some blank music paper, wafers; a box for botanizing; a thermometer; two little bottles with spirits for preserving insects; a bag for picking up stones; two maps of Scotland—Ainsley's small one, and Sayer's; letters of recommendation. ... I found my bags at first an encumbrance, but became afterwards more reconciled to them. They are to a saddle what pockets are to a coat, and who objects to wearing pockets? But they were wetted the first day, and stained their contents; this will make me more careful in future.

The two-month tour was a very extensive one: up to John o'Groats in the far north-east and over to the island of Mull in the far west. And the going was really tough. During a certain section, Young's horse "was obliged to creep up or to slide down steep hills, to push his way through rocks, or to step delicately over boggy ground"; at one point, it sank so far into a bog that its rider's feet touched the ground, "but we soon got out."

The romantic adventures of Don Quixote and Orlando Furioso seem to have helped the solitary Young to keep his morale up, for he writes:

> To lose one's way in a dark night, to have to pass through rocks and bogs, to ford deep waters, to cross steep mountains, to stand long in waiting for an asylum at a late hour in a miserable hut; to be prepared for deranged accoutrements, a lame horse, his shoes loose, his back galled, his spirits flagging; and again after a short time to be welcomed with as much hospitality, and entertained with as much splendor, as any lord of a castle could receive a knight-errant: to be at ease from every care and in the enjoyment of every amusement that men of sense and women of elegance can afford: all these vicissitudes exercise the same qualities, require the same virtues, and excite the same emotions as the obsolete chivalrous tales of fabulous ages.

He was not exaggerating about Scottish Highland hospitality. His many letters of recommendation—from, for example, Dalzel to a fellow professor at St Andrews, and from the mother of the duke of Richmond to the duke and duchess of Gordon at Gordon Castle—opened all doors to Young. His activities ranged from inspecting a "rich store of mathematical and philosophical apparatus" at Aberdeen University and the objects turned by the duke of Gordon on the lathes of Gordon Castle, to accompanying a stag hunt and dancing energetic Scottish reels. All engaged his detailed attention, but perhaps the highlight of the tour was the company of young ladies. Peacock calls Young "passionately fond of female society", and his Scottish journal proves the truth of this. At Gordon Castle, he brought out some of his notes for the ladies and "read them some of my extracts in verse, and I thought the better of my selection when I found that Lady Louisa had some of them by heart". While at Inverary Castle, staying with the duke of Argyll, he notes:

> I was showing Lady C. some of my sketches; she begged to see my notes, and I showed the greatest part of them. All the family are musical; the ladies sing admirably; cards and the fine piano occupied the evening. After supper, besides other songs, I heard a most beautiful canzonet [a kind of madrigal] by Jackson, beginning "Love in Thine Eyes". It was twelve o'clock when we retired. After breakfast I took my leave; not without regretting that I had so little time to observe the beauties of Inverary. Lady Charlotte is handsomer than Lady

Augusta, she sings better, but she has less good sense, and less sweetness; an innocent giddiness sometimes gives her the appearance of a little affectation; she is to Lady Augusta what Venus is to Minerva; I suppose she wishes for no more. Both are goddesses.

Back in Edinburgh on 6 August, after a short stop to wind up his personal arrangements and see some friends, Young rode south to England, stopping in the Lake District, in Liverpool (to see his friend Bostock) and in north Wales. He sold his horse in Birmingham and proceeded by coach to London, dropping off on the way for a day with Edmund Burke and arriving in early September at his great-uncle's house, where he was anxiously awaited by Brocklesby.

Just before leaving Edinburgh, Young had written to his Quaker mother, outlining his plans whilst also trying to reassure her that he was not going astray (despite some reports she perhaps had heard from Quakers in Edinburgh). He mentioned that in the fall he would proceed either to Leiden in the Netherlands, or more probably Göttingen in Germany, and after graduating, travel on to Vienna, and then to Pavia, Rome and Naples—"after this I must be regulated by the state of politics", in other words Britain's war with France. Physicians, he said, should cultivate their minds more widely than through science alone:

> I think I cannot better spend the next two years of my life, than in attending (at the same time I continue my scientific pursuits under the most eminent professors in different parts of Europe) to the various forms into which the customs and habits of different countries have moulded the human mind; in imitating what is laudable, and in avoiding what is culpable, and in exerting myself to gain the acquaintance and friendship of the virtuous and learned.

Göttingen University, where Young arrived in late October, already had a high reputation, though it was small and had been founded only half a century before, in 1737, by George, elector of Hanover and afterwards George II of England. Just over a hundred of its roughly eight hundred students were studying medicine. For Young, the greatest attraction was its library. "I am within a hundred yards of the second library in Europe, and can have any book I wish to consult on sending for it; this is the chief reason for my desiring to graduate here," he told his mother in another letter—and it is clear from his dissertation at Göttingen that he made full use of the library books and manuscripts in many languages.

By a pleasing coincidence, he lived in a building that later became an institute for physics: a subject in which Göttingen would excel in the nineteenth and twentieth centuries (when it became a founding center of quantum mechanics). There, in 1833, Karl Friedrich Gauss and Wilhelm Weber invented an electromagnetic telegraph. When the coincidence was discovered in 1911, a tablet commemorating Young, too, was placed on the building next to the tablet in memory of Gauss and Weber.

Apart from attending the medical lectures, he was soon taking lessons in music, drawing and horsemanship; at this period, Göttingen was considered the first school of horsemanship in Europe. Even for Young, it was difficult to follow everything in German initially, but he soon became fluent, partly because he and his fellow English students made a pact to speak only in German on pain of forfeiting twopence every half hour. He mentioned this pact in a letter to Bostock, saying that "all of a sudden two Scotchmen came in upon us, and one of them speaking not a syllable but English, and obliging me and Colhoun out of civility to incur frequent penalties; the other two, who are more solitary beings, have escaped." Then he added: "We are not riotous nor absurd, as Englishmen in foreign countries generally are said to be, but it will require some influence to restrain young H—'s propensity to drinking."

English insularity was largely foreign to Young's character, unlike Brocklesby, who had worked in Germany as an army physician decades previously. His great-nephew found himself anxious, on some occasions, to defend German medicine and the German people in general. "The English physicians are quoted as familiarly here as they are at home; the Germans know what London contains better than many of our own countrymen; but the reverse does not hold good; the German authors are very imperfectly or not at all known in Britain," he wrote to Brocklesby. Thus, "the science here has one advantage—that the doctrines of both countries are well known here, while the English attend little to any opinions but those of their own country."

But he was unfavorably struck by the difference in hospitality between the Edinburgh and Göttingen professors. Although he met and was on good terms with a number of them, including the director of the library and eminent classicist C. G. Heyne, relations remained relatively formal. To his friend Dalzel, the professor of Greek in Edinburgh, Young wrote:

> The little disposition to social intercourse that is found in most of the professors, precludes almost the possibility of an intimacy with any of the rank of a student. I have never, for instance, been more in Heyne's house than about four times for a quarter of an hour of a morning, and do not know immediately from him how his literary undertaking proceeds, so that I can say nothing of his Homer.

To Brocklesby, he commented: "on the whole, one must be content to be in Göttingen a mere student".

As early as April 1796, Young felt ready for an examination by the medical faculty, in preparation for receiving his degree after submitting a dissertation. For four or five hours he and the four examiners sat around a table, "well furnished with cakes, sweetmeats, and wine, which helped to pass the time agreeably". The questioning, according to Young, was astute, "but the professors were not very severe in exacting accurate answers." Presumably, his competence was quickly obvious to them. He passed on 30 April.

His dissertation was printed and circulated in June. Written in Latin and dedicated to "Ricardo Brocklesby Societatis Regiae Londinensis" ("Richard Brocklesby of the Royal Society of London"), it concerned the organs of the human voice, and put forward an alphabet of 47 letters designed to express, alone and in combination, the entire gamut of sounds that these organs are capable of producing. This 'universal' alphabet would be useful, said its author, for writing down the unwritten tribal languages of Africa and America. Thus, his dissertation subject united anatomy with two of Young's current and future interests: language and sound. It was highly praised by the faculty, publicly defended by Young in July in the customary manner, and then, after he had read something like a prayer (the normal oath was waived, given his Quaker background), he was "married to Hygeia, and created doctor of physic, surgery, and midwifery."

Brocklesby was delighted by the thesis, naturally enough, while noting that its true quality would not be widely appreciated. Young replied: "You say my thesis is caviar to the general; but do not you think people have a greater respect for anything out of the common way? Perhaps indeed few people will give it attention enough either to like or dislike it, yet I do not know that I have any reason to avoid distributing it among my friends." Then he made a significant statement, given the lifelong criticism he would suffer for being over-concise: "It seems a fatality that almost everything I do or produce should be termed stiff: in this case it may arise from my having been obliged to treat the subject in a short compass."

It was now time for Young to begin his travels in Europe. In May, during a break in the lectures, he had already made a trip to the Harz Mountains with two English fellow-students (one of them being the son of the potter Josiah Wedgwood), where they descended two of the deepest mines and ascended the legendary Brocken. Now, in late July, at last free of commitments to Göttingen, he set off on an extended tour of the independent German states, having abandoned his longer-term plan to winter in Vienna and then visit Italy because of Napoleon's Italian victories in mid-1796. (As Young was amused to note in his autobiographical sketch, Napoleon delayed his planned tour of Italy for 25 years, until he finally got there in 1821, "while that great conqueror was dying at St Helena".) The tour took him to Brunswick, Gotha, Weimar, Jena, Leipzig, Dresden and finally Berlin.

As in the Scottish Highlands, he mixed with the best society, speaking variously German, French, Italian and English. For example, while at the court in Weimar he met the critic and philosopher Johann von Herder, Goethe's friend, whose son had been at Göttingen with Young. Herder was curious about his thesis on the human voice, so Young wrote out for him "a little specimen of the manner in which I would describe the pronunciation of the most current European languages". He noted that though Herder was very well versed in English poetry, he did not speak English. "We had a little debate on the subject of rhyme, which he would reject altogether." And while at Leipzig he met a Silesian count, who gave him a seat in his chaise upon condition that Young paid for an extra horse. "We traveled in the finest moonshine, and in the morning reached the Elbe at Meissen, one of the finest river views which I have ever seen"—and from there went on to Dresden.

In that great city of culture, he stayed a month and took the opportunity of its vast collections of fine art to study at length the choicest works of Italian art, which had been denied to him by Napoleon's armies. Although drawing was one of Young's lesser skills, it always keenly interested him. Berlin, by contrast, seems to have been a bit of a letdown, despite his dining twice with Lord Elgin, the British ambassador. "I have not formed many connections in this place, neither have I seen any great curiosities; indeed I am glad to have a little respite from the perpetual pursuit of novelties, which seldom equal expectation," he wrote to Brocklesby after three weeks there. But he stayed in Berlin over a month and a half, before sailing back from Cuxhaven near Hamburg on the packet to England—a rough, eight-day journey overshadowed by sea sickness and the threat of attack by privateers—in early February 1797.

Young had been away from home for well over two years, and the experience had altered him. In a long analysis of the differences between Germany and Britain he wrote at the time of his return, he remarked:

> [T]here are more learned men in Germany than in England, but we have, and ever have had, some individuals in many branches who are almost unequaled. Latin is much better understood in Germany, Greek but little; commercial men speak French and often English. In mathematics and in chemistry the Germans are making rapid advances; as painters they copy better than the English, but have perhaps less invention; in engraving the English confessedly excel them; and the Germans still more decidedly bear away the palm in music, in which they rival the Italians.

It is not hard to hear in this the voice of the future foreign secretary of the Royal Society. Even two centuries ago, science was more international, by its very nature, than other branches of knowledge. Young would be among the first scientists fully to embody that internationalism.

Chapter 4

'Phenomenon' Young

"I am at present a good deal employed on the subject of the slight synoptic sketch at the end of my thesis, the definition and classification of the various sounds of all the languages that I can gain a knowledge of; and have of late been diverging a little into the physical and mathematical theory of sound in general. I fancy I have made some singular observations on vibrating strings, and I mean to pursue my experiments."

Young, letter to Andrew Dalzel, 1797

After four years' training in medicine, from 1792 to 1796, in London, Edinburgh and Göttingen, Young hoped to be eligible to begin practicing as a physician in London on his return from Germany in February 1797. But he was in for a major disappointment. Earlier, he had been advised by knowledgeable physician friends that two years of university study would suffice for him to qualify as a licentiate of the College of Physicians, which existed to regulate "all persons practicing physic within the city of London, and a circuit of seven miles round it" (according to its original royal charter of 1518). However, the statute of the college had very recently been changed and now required licentiate physicians to study for two years at the same university. A year spent at Edinburgh and a second year at Göttingen were no longer deemed sufficient to qualify as a licentiate. Young, who was now almost 24, was therefore obliged to return to university, much against his inclinations.

He chose to spend the requisite years at Cambridge University, in order to obtain the degree of bachelor of medicine (M.B.). It made sense to go there for professional reasons because Oxford and Cambridge men had a stranglehold over the College of Physicians. To become a Fellow of the Royal College of Physicians, an F.R.C.P. (as his great-uncle was), it was necessary, in practice if not in theory, to hold an Oxford or Cambridge doctoral degree (M.D.). Physicians possessing such a doctorate were automatically candidates for fellowship, and became fellows after a year's probation. Physicians with doctorates from other universities, though permitted to practice as licentiates, were eligible as candidates for fellowship only after a period of seven to ten years on the nomination of the president of the college. As a result, some of the leading eighteenth-century physicians, for instance John Fothergill, William Hunter and John Coakley Lettsom, were not fellows of the College of Physicians (despite being fellows of the Royal Society). This requirement created much resentment, but the Oxford and Cambridge graduates running the College of Physicians had a vested interest in protecting their privileges in London's medical world and had long been resistant to reform.

Young's decision made sense for personal reasons as well. The master of Emmanuel College in Cambridge, Richard Farmer, a Shakespeare scholar, was an intimate of Dr Brocklesby, and through him Farmer had become acquainted with Young too; indeed, Farmer had been one of the sponsors of Young's fellowship of the Royal Society in 1794. Brocklesby suggested that his great-nephew go to Farmer's college. Unfortunately, Farmer died very soon after, but it was to Emmanuel that Young proceeded in March 1797, as a so-called fellow commoner—that is, a student with the right to dine with the college fellows—not as a mere undergraduate. He would reside there, at least formally, until the end of 1799, while of course spending periods of time in London, especially after he inherited Brocklesby's Norfolk Street house upon his great-uncle's death at the end of 1797.

In going to Cambridge University, Young finally had to cut his links with the Quakers. In 1797, in fact as late as 1871, every candidate for a bachelor's degree at Cambridge had to declare that he was "bona fide a member of the Church of England as by law established". Dissenters were excluded. In being admitted to Cambridge, Young therefore became a de facto member of the Church of England, whatever the true state of his religious views. In February 1798, he was formally disowned by the Quakers, though the reason given was not Cambridge-related but that he had "attended places of public diversion"; in other words, theaters and dances. He had been interviewed by a Quaker deputation from the Westminster meeting house and shown no remorse for his conduct; rather, "by his own acknowledgment [he was] estranged from us in principle and practice", notes the entry on Young in the *Dictionary of Quaker Biography*. (His friend Hudson Gurney was disowned in 1803 for "making [a] contribution to a fund for military purposes at a time of danger of French invasion".) There is no reference to this interview by Young himself in his surviving writings, and his biographer Peacock, surprisingly for a divine, remains silent on the subject. From now on, Quakerism would hardly ever be mentioned by Young; and, astonishingly, he would make no mention at all of his parents—at least judging from his letters to Gurney and family members quoted by Peacock—despite the fact that his mother lived well into her sixties, dying in 1811, and his father reached the age of 73, dying in 1819. Young would never abuse the Quakers and their religion, but neither would he praise them.

"The foolish laws of the College in London are perplexed and ill understood; but I must now make the best of Cambridge", Young wrote to Andrew Dalzel in April, a month after his arrival, presumably in some exasperation. The sole comment he makes on Cambridge in his autobiographical sketch written three decades later is: "he did not think it necessary to attend the lectures of any kind, in subjects which he had before sufficiently studied." Nor was he particularly interested in diverting himself with traditional college and university entertainments (whatever the strait-laced Quakers might think), having perhaps had his fill of them in Edinburgh and Göttingen.

Most of his time at Cambridge would be spent in solitary reading, writing, and doing experiments in physics—as opposed to physic—in his college rooms. His reading about the human voice, done for his dissertation at Göttingen, inspired him to go more deeply into the subject, encouraged by Brocklesby and his friend, the eminent physician William Heberden Sr. As Young noted a few years later:

> When I began the outline of an essay on the human voice, I found myself at a loss for a perfect conception of what sound was, and during the three years that I passed at Emmanuel College, Cambridge, I collected all the information relating to it that I could procure from books, and I made a variety of original experiments on sounds of all kinds, and on the motions of fluids in general.

The results would be published only after he left Cambridge, in a series of striking papers read to the Royal Society from 1800 onwards, in his lectures at the Royal Institution, and in his great book of these lectures published in 1807; and so we shall leave them to later chapters. Overall, the period when Young lived in Cambridge has a whiff of Newton's reclusive brilliance about it—though Young was a far more sociable man than his great predecessor in natural philosophy. Newton spent 35 years in Cambridge, but apparently wrote to no one at all there after he left the city in 1696 before his death some thirty years later in 1727; such behavior would have been wholly inconceivable for Young.

Peacock and Wood, Young's biographers, being Cambridge men to the core who dedicated their entire working lives to the university, appear somewhat embarrassed about Young's indifference to Cambridge. They badly want Cambridge to matter to Young—but in truth it did not. He was only too aware that it had no medical school and very few medical lectures worth the notice, as compared with London and Edinburgh. In fact, its intellectual life as a whole was only just beginning to recover from a century-long period of mediocrity, as was also the case in eighteenth-century Oxford. "The professors of the university seldom performed any of their supposed functions," wrote G. M. Trevelyan in his *Social History of England*. "No lecture was delivered by any regius professor of modern history (at Cambridge) between 1725 and 1773; the third and most scandalous of the holders of that chair died in 1768 from a fall while riding home drunk from his vicarage at Over." The surgeon John Hunter, despatched to Oxford by his brother William in 1755 presumably to acquire some much-needed polish for a career as a physician, lasted a mere two months there before storming back to London, having refused to "stuff Latin and Greek". Young certainly had no need of Cambridge in that particular department, and must soon have realized that he had relatively little to learn in other departments too. "I am ashamed to find how much the foreign mathematicians for these forty years have surpassed the English in the higher branches of the sciences," he wrote to Dalzel in mid-1798. Of a revered Cambridge mathematician, Robert Smith, master of Trinity College in the middle of the century and founder of the Smith's prizes in mathematics, he added caustically: "I think [he] has only been admired because few would trouble themselves to wade through so much affected obscurity."

Young's personality was now fully formed, and it is tempting to see it reflected in the heartfelt way he much later portrayed his friend Richard Porson, the classical scholar from a lowly social background who gave up his fellowship of Trinity in 1791, after refusing to be ordained in the Church of England. Young writes of Porson:

> [H]aving learned "to know how little can be known", it is not surpris-
> ing that he found himself "without a second and without a judge",
> and that he was unwilling to affect a community of sentiment and an
> interchange of approbation with those whose acquirements and
> opinions he felt that he had a right to despise. It might have been
> wiser in some instances to conceal this feeling; but, on the other
> hand, he had perhaps occasion for something of the habit of retreat-
> ing into his conscious dignity, from his deficiency in those general
> powers of ephemeral conversation, which are so valuable in mixed
> societies: for, with all his learning and all his memory, he was by no
> means prominent as a talker. He had neither the inclination nor the
> qualifications to be a fascinating story-teller, or to become habitually
> a parasite at the tables of the affluent; but he was the delight of a lim-
> ited circle of chosen friends, possessing talent enough to appreciate
> his merits, and to profit by the information that he afforded them.

The chief source of information about Young at Cambridge is like a counterpoint to this putative self-portrait. It is a most revealing and vigor-ous sketch of Young written by a tutor at Emmanuel College, presumably in the 1830s or 40s, at the request of Young's biographer Peacock (who himself was a tutor at Trinity). Revealing, that is, about both Young and about the Cambridge of his day. Peacock does not name his source but comments, a shade defensively, before quoting the sketch at length, that the writer "was a man of great energy of character and of very acute observa-tion, but possessed of no great learning". Whoever he was, he clearly did not warm to Young as a man, and though he knew of his scientific achievements, he was not willing to recognize their importance, even after Young's death. Rather, one cannot avoid feeling that the Emmanuel tutor seems to have seen Young as, at bottom, a failed Cambridge man deficient in college spirit.

The sketch begins: "When the master introduced Young to his tutors, he jocularly said: 'I have brought you a pupil qualified to read lectures to his tutors.' This, however, as might be concluded, he did not attempt, and the forbearance was mutual; he was never required to attend the common duties of the college."

First, it deals flatteringly with Young's classical scholarship:

He had a high character for classical learning before he came to Cambridge; but I believe he did not pursue his classical studies in the latter part of his life—he seldom spoke of them; but I remember his meeting Dr Parr in the college combination room, and when the doctor had made, as was not unusual with him, some dogmatical observation on a point of scholarship, Young said firmly: "Bentley, sir, was of a different opinion"; immediately quoting his authority, and showing his intimate knowledge of the subject. Parr said nothing; but, when Dr Young retired, asked who he was, and though he did not seem to have heard his name before, he said, "A smart young man that."

He had a great talent for Greek verse; and, on one occasion, I remember a young lady had written on the walls of the summer-house in the garden the following lines:

> *Where are those hours on airy pinions borne*
>
> > *That brought to every guiltless wish success?*
>
> *When pleasure gladdened each succeeding morn,*
>
> > *And every ev'ning closed with dreams of peace?*

On the next morning appeared a translation in Greek elegiacs, written under them, in Young's beautiful characters. It may be here mentioned, that when his mode of writing Greek was laid before Porson, he said, that if he had seen it before he would have adopted it.

Then the writer turns to Young's knowledge of the sciences and becomes more critical:

> The views, objects, character, and acquirements of our mathematicians were very different then to what they are now, and Young, who was certainly beforehand with the world, perceived their defects. Certain it is, that he looked down upon the science and would not cultivate the acquaintance, of any of our philosophers. Wood's books I have heard him speak of with approbation, but Vince he treated with contempt, and he afterwards returned the compliment. I recollect once asking Vince his opinion of Young: he said he knew nothing correctly. "What can you think," says he, "of a man writing upon mechanics, who does not know the principle of a coach wheel." This alludes to a mistake of Dr Young's on this subject in his *Natural Philosophy*.

As for Young's knowledge of contemporary life and letters, and his sense of humor, the sketch portrays him as comically out of touch:

> He did not seem even to have heard the names of most of our poets, or literary characters in the last century, and hardly ever spoke of English literature. I remember having invited him to meet at dinner Mr Whiter, of Clare Hall, who, though an admirable scholar, was a wit and a *bon-vivant*, while Young took no delight in the pleasures of the table, and never could make a joke or understand one. Whiter quoted something from the *Oxford Sausage*, and when our philosopher betrayed his ignorance of the existence of such a work, with his total inability to taste or relish the allusion, it was almost painful to witness the ridicule which he was obliged to sustain; but to do him justice, he did sustain it with perfect good humor.

Then, in the most vivid section of the sketch, the tutor grapples with Young's personal presence and behavior, which he clearly found baffling and thoroughly un-Cantabrigian:

> He never obtruded his various learning in conversation; but if appealed to on the most difficult subject, he answered in a quick, flippant, decisive way, as if he was speaking of the most easy; and in this mode of talking he differed from all the clever men that I ever

saw. His reply never seemed to cost him an effort, and he did not appear to think there was any credit in being able to make it. He did not assert any superiority, or seem to suppose he possessed it; but spoke as if he took it for granted that we all understood the matter as well as he did. He never spoke in praise of any of the writers of the day, even in his own peculiar department, and could not be persuaded to discuss their merits. He was never personal; he would speak of knowledge in itself, of what was known or what might be known, but never of himself or any other, as having discovered anything, or as likely to do so.

His language was correct, his utterance rapid, and his sentences, though without any affectation, never left unfinished. But his words were not those in familiar use, and the arrangement of his ideas seldom the same as those he conversed with. He was, therefore, worse calculated than any man I ever knew for the communication of knowledge. I remember our once asking him to answer an objection to Huygens's theory of light, which he preferred to Newton's, and, though there were many very competent persons present, he attempted in vain. ...

In his manners he had something of the stiffness of the Quakers remaining; and though he never said or did a rude thing, he never made use of any of the forms of politeness. Not that he avoided them through affectation; his behavior was natural without timidity, and easy without boldness. He rarely associated with the young men of the college, who called him, with a mixture of derision and respect, "Phenomenon Young"; but he lived on familiar terms with the fellows in the common-room. He had few friends of his own age or pursuits in the university; and not having been introduced to many of those who were distinguished either by their situation or talent, he did not seek their society, nor did they seek him: they did not like to admit the superiority of any one *in statu pupillari*, and he would not converse with any one but as an equal.

It was difficult to say how he employed himself; he read little, and though he had access to the college and university libraries, he was seldom seen in them. There were no books piled on his floor, no papers scattered on his table, and his room had all the appearance of belonging to an idle man. I once found him blowing smoke through long tubes, and I afterwards saw a representation of the effect in the *Transactions* of the Royal Society to illustrate one of his papers on sound; but he was not in the habit of making experiments. He walked little, and rode less, but having learnt to ride the great horse abroad, he used to pace round Parker's Piece on a hackney; he once made an attempt to follow the hounds, but a severe fall prevented any future exhibition.

At the end of the sketch, the writer attempts a kind of summing up of Young:

> He seldom gave an opinion, and never volunteered one. He never laid down the law like other learned doctors, or uttered apothegms, or sayings to be remembered. Indeed, like most mathematicians, (though we hear of abstract mathematics,) he never seemed to think abstractedly. A philosophical fact, a difficult calculation, an ingenious instrument, or a new invention, would engage his attention; but he never spoke of morals, of metaphysics, or of religion. Of the last I never heard him say a word, nothing in favor of any sect, or in opposition to any doctrine; at the same time, no sceptical doubt, no loose assertion, no idle scoff ever escaped him.

By now it will be obvious that this sketch, valuable though it is biographically speaking, is a superficial portrait of a man who did not by nature easily open himself to others (as Young himself had written of Porson). The criticism of Young's horse riding by the Emmanuel tutor is a particular give-away; no one who had ridden a horse alone through the Scottish Highlands for two months could possibly have been thrown by a little hunting in flat Cambridgeshire. Yet the sketch probably did represent the views of most of those who encountered Young at Cambridge, not to speak of some academics today. Young made mediocrities uneasy, and to cover their unease, they belittled and ridiculed him.

There were, however, a number of exceptions among Young's Cambridge contemporaries, as the touchy Peacock was at pains to point out: he gives a substantial list of them by name and qualification. It would definitely be wrong to give the impression that Young was socially isolated at Cambridge. But his relationships were not, in the main, college-based;

Young picked and chose his acquaintances and friends from across the university and across disciplines, including physic. It was a *modus vivendi* that he would pursue for the rest of his life.

Even in Emmanuel College itself, the records of Young, scanty as they are, show that he was by no means unpopular. He was not naturally collegial, but he enjoyed a like-minded group. After only six months, he was elected president of the Emmanuel Parlour (the common-room), in November 1797—a "signal honor", says Alex Wood, who was a fellow of Emmanuel in the twentieth century. A college book records the bets won and lost by the fellows. Young successfully betted on the angle subtended at the sun by the earth's semi-diameter and, twice on the same night, betted "that Young does not produce thirty pins the wires of which occupy less space than an inch", and "that Young does not draw with a pen one hundred lines in the space of an inch". His success in this second bet was a signal honor for his skill as a calligrapher. Not all his bets were successful, though. One that was left undecided, and then given against him after he had left Emmanuel, was that "Young will produce a pamphlet or paper on the theory of sound more satisfactory than any thing that has already appeared, before he takes his bachelor's degree" (which would formally be in 1803). This plainly required a judgment call, and one wonders how many of the fellows of Emmanuel were really competent to make such a judgment.

During that first year at Cambridge, Young had written a number of approving letters to his great-uncle in London, and Brocklesby was no doubt gratified to hear that Young had been elected president of the common-room. About a month later, on 13 December 1797, just after retiring to bed, Brocklesby suddenly died. His great-nephew had been with him at dinner and found him in good spirits, having returned from Cambridge that very day.

Brocklesby's will bequeathed the greater half of his fortune, including his Irish estates, to his other nephew, Mr Beeby. About half of the remainder went to Young, consisting of the house and furniture in Norfolk Street, Brocklesby's library, his prints, and a small but choice collection of pictures chiefly selected by his friend Sir Joshua Reynolds, along with £10,000 in money (today worth about fifty times that figure).

"Young had just reason for regarding with affection the memory of this kind and liberal relative," Peacock comments on Brocklesby. But he felt obliged to add, in the interests of biographical honesty:

> It is quite true that even the kindest actions of this excellent relative were not altogether unmixed with some root of bitterness, such as dependence of every kind, except that of a child upon a parent, is apt to bring along with it: he was somewhat querulous in his temper, and somewhat exacting in his claims to respect and deference; though liberal in great things, he was somewhat parsimonious in small; and though generally judicious in the course which he recommended Young to pursue, he was sometimes rather unreasonably suspicious when his wishes (though often very obscurely intimated) were not fully carried out.

At least a part of Young's mind must have been relieved to see old Brocklesby go, though naturally he put nothing on the subject on paper. His death came at the perfect moment for his great-nephew, when he was no longer in need of help or guidance, and liberated him from immediate financial worries. Young transferred some of his own books from Emmanuel College to his great-uncle's residence in London, imported some pictures from there into Emmanuel to make his college rooms cheerful, and acquired a horse and a servant. In his autobiographical sketch, he commented that the inheritance was "enough to afford him a sufficient degree of temporary affluence or even scanty subsistence, without annihilating or much weakening the motives for application to a profession." When he returned to London from Cambridge for good two years later, the world would be truly his oyster.

Chapter 5

Physician of Vision

"His pursuits, diversified as they were, had all originated in the first instance from the study of physic: the eye and the ear led him to the consideration of sound and of light."

Young, *"Autobiographical sketch", 1826/27*

The year 1800 marked the beginning of a new century and the most scientifically significant phase of Young's life, during which he published the work that would change physics and physiology. But it did not launch his career as a physician. Although he was now permitted to practice medicine in London, he was not yet the holder of a Cambridge M.B. degree, could not become an M.D. under Cambridge regulations until five years after that, and would only then be eligible to become a Fellow of the Royal College of Physicians (F.R.C.P.)—these titles were conferred on him in 1803, 1808 and 1809, respectively, considerably later than for most physicians. In the meantime, he would need to begin looking for a suitable position in a hospital as a physician and building up a private practice, since he could expect to inherit no more than a few of his great-uncle's patients. (The most famous of them, Edmund Burke, had died the previous year, a few months before Brocklesby.)

Perhaps this is why Young decided to sell Brocklesby's house in Norfolk Street off fashionable Park Lane opposite Hyde Park, and find himself a new address in 1801. Young left no clue as to his motives in moving house soon after his return to London, but the wish to live more modestly than had Brocklesby, at least until he established himself as a physician, would have been natural and prudent. He probably also wanted to get away, though not far away, from his great-uncle's shade and the many memories—not all of them pleasant ones—that would have crowded him at 10 Norfolk Street. And perhaps he also sensed that he should locate himself nearer to the emerging center of the medical profession around Harley Street, an address that within a few decades would become synonymous with top-ranking physicians and surgeons. All these reasons would explain his choice of Welbeck Street, close to Harley Street, where he would live for the next 25 years.

He might well have chosen instead to look for a house in the area around Soho and Leicester Square, where Newton, Hogarth, Reynolds and John Hunter had lived, and where the Hunterian school of anatomy was located in Great Windmill Street. Such an address was certainly prestigious, the atmosphere was intellectual and artistic, and the streets were unquestionably lively, being close to the fashionable Strand, to theaters and to sinful Covent Garden. But maybe the area would have seemed too rakish for an up-and-coming physician, despite Young's sympathy for Georgian free enterprise (as epitomized by the Hunterian school). He must have preferred somewhere quieter and more sedate in the residential streets that had recently been built north of Oxford Street and south of the New Road, the route laid out in 1756-57 as the northern boundary of London in the late eighteenth century (now the Marylebone, Euston and Pentonville Roads). In this period, London, having expanded westwards and created the West End earlier in the century, shot northwards—so much so that in 1791 Horace Walpole, the writer and politician, who lived in the West End (in Berkeley Square) to the south of Oxford Street, complained that London's expansion was killing the sedan-chair trade, "for Hercules and Atlas could not carry anybody from one end of this enormous capital to the other".

These new north London streets and squares—with some substantial mansions built by well-known architects like Robert Adam—were constructed on the fields of the Harley-Cavendish estate, land belonging to Edward Harley, the second earl of Oxford, who had formed a shrewd marriage alliance with the Cavendish family. Starting from the nucleus of Cavendish Square, streets such as Harley Street, Portland Place, Bentinck Street (the latter two named after William Bentinck, the second duke of

Portland, who married a Cavendish daughter), and Welbeck Street (named after Welbeck Abbey, the Portland family home), were developed. Among the new residents of the area were the Maxwell family of Cavendish Square, one of whose daughters Young would soon marry; and, a little before Young, Edward Gibbon, author of *The History of the Decline and Fall of the Roman Empire*, much of which was written at 7 Bentinck Street, just around the corner from Young's future house, in 1773-83. The bachelor Gibbon sold his own family estate when his father died so as to buy this desirable new town house in Bentinck Street, and wrote smugly of his good fortune:

> I had now attained the solid comforts of life, a convenient well-furnished house, a domestic table, half a dozen chosen servants, my own carriage, and all those decent luxuries whose value is the more sensibly felt the longer they are enjoyed ... To a lover of books the shops and sales in London present irresistible temptations ... By my own choice I passed in town the greatest part of the year.

It is unlikely that Young lived in luxurious style at 48 Welbeck Street, though we cannot be sure about this since he seems never to have described his domestic arrangements in any detail, except for the fact that he kept a horse and servants. But like Gibbon, Young certainly aspired to elegance and high social status, appreciated comfort, and relished London society and metropolitan booksellers, though without any ostentatious display. What mattered most to him was the freedom his considerable wealth gave him to work on the subjects that appealed to his mind. No sooner had he moved into the new house than he was apologizing in a letter to Andrew Dalzel, dated 27 June 1801, for the delay in replying to him caused by "the confusion of furnishing and entering upon a house." There must have been a great many books and some scientific equipment to be transported, too, along with the furniture and pictures, by horse-drawn conveyance from Norfolk Street. He continued:

> I am at present employed in some further optical investigations, which, I imagine, will be considered as more important than any of my former attempts, as I think they will establish almost incontrovertibly the undulatory system of light, and extend it to the explanation

of an immense variety of phenomena. I have also some prospect of being in a situation which will enable and require me to devote more time to the pursuit of natural philosophy than I should otherwise think consistent with the profession of physic, but the idea is yet only in embryo.

The "situation" Young referred to was the tempting offer of the post of professor of natural philosophy at the Royal Institution, which would keep him extremely busy with (entirely non-medical) lectures in 1802-03. We shall come to this in the next chapter. As for the "further optical investigations" connected with the wave theory of light (and its comparison with sound), these too fully deserve their own separate chapter, after the Royal Institution lectures, since they did not come to full fruition until the end of 1803, and also because they are fundamentally physics and therefore distinct from Young's simultaneous physiological work on the human eye and color vision in 1800-01, which grew out of his knowledge as a physician.

"On the mechanism of the eye", Young's lengthy and highly detailed paper read to the Royal Society in November 1800, and published in 1801, was called, more than a century later, a "masterly monograph" containing "Young's greatest and most original contributions to science". Admittedly the comment came from an ophthalmologist and physiologist, a distinguished one, Sir John Parsons, who was perhaps partisan; yet this particular paper has been so widely admired for its experimental ingenuity and the clarity of its deductions that we shall quote from it at length. It was here that Young finally established the process at work in the accommodation of the eye, and also defined and measured astigmatism for the first time.

Bearing in mind the contested nature of the field—the differing ideas of Kepler and Descartes about accommodation, Porterfield's experiments on couched eyes, Young's own first paper of 1793, Hunter's claim to priority, the rumor of plagiarism against Young, the contrary paper by Home published in 1795, Young's subsequent withdrawal of his thesis about the 'muscular' lens and Monro's skepticism about both Young's paper and Home's when Young was a student at Edinburgh, not to mention even more recent evidence from others against the lens as the site of accommodation—Young felt the need for a semi-historical and cautionary preamble. His paper therefore begins:

In the year 1793, I had the honor of laying before the Royal Society, some observations on the faculty by which the eye accommodates itself to the perception of objects at different distances. The opinion which I then entertained, although it had never been placed exactly in the same light, was neither so new, nor so much forgotten, as was supposed by myself, and by most of those with whom I had any intercourse on the subject. Mr Hunter, who had long before formed a similar opinion, was still less aware of having been anticipated in it, and was engaged, at the time of his death, in an investigation of the facts relative to it; an investigation for which, as far as physiology was concerned, he was undoubtedly well qualified. Mr Home, with the assistance of Mr Ramsden, whose recent loss this society cannot but lament, continued the enquiry which Mr Hunter had begun; and the results of his experiments appeared very satisfactorily to confute the hypothesis of the muscularity of the lens. I therefore thought it incumbent on me to take the earliest opportunity of testifying my persuasion of the justice of Mr Home's conclusions, which I accordingly mentioned in a dissertation published at Göttingen in 1796, and also in an essay presented last year to this society. About three months ago, I was induced to resume the subject, by perusing Dr Porterfield's paper on the internal motions of the eye; and I have very unexpectedly made some observations, which I think I may venture to say, appear to be finally conclusive in favor of my former opinion, as far as that opinion attributed to the lens a power of changing its figure. At the same time, I must remark, that every person who has been engaged in experiments of this nature, will be aware of the extreme delicacy and precaution requisite, both in conducting them, and in drawing inferences from them; and will also readily allow, that no apology is necessary for the fallacies which have misled many others, as well as myself, in the application of those experiments to optical and physiological determinations.

To understand the paper itself, we need first to review the basic structure of the human eye, which was understood in Young's day, except for the function of the ciliary body. As shown in Figure 5.1, the front of the eye has an outer transparent cover, the cornea, which merges into the non-transparent white of the eye, the outer covering of the eyeball, known as the sclera. The cornea encloses a space filled with a fluid called the aqueous humor. Behind this is the iris diaphragm, which controls the size of the pupil, as in a camera. And behind the iris is the crystalline lens surrounded by the ciliary muscles, the inner space of the eye filled with a second fluid known as vitreous humor, and the light-sensitive inner coating of the inside of the eye, the retina, which is connected to the optic nerve and through it to the brain.

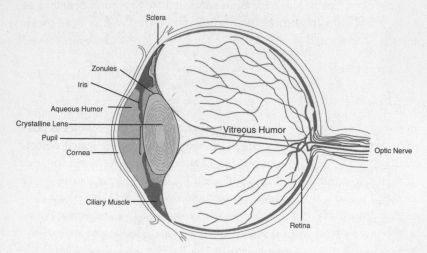

Figure 5.1   *Cross-section of the human eye. The ciliary muscles and zonules were unknown to Young, who assumed the lens was muscular.*

Rays of light from a point on an object, traveling through the air-cornea interface into the aqueous humor, are bent by refraction (compare the apparent bending of a pencil dipped in a cup of water). Passing through the pupil, the rays are then further bent by refraction in the crystalline lens. About two thirds of the eye's bending takes place at the air-cornea interface, the rest in the lens. If the eye is correctly focused for the distance of the object from it, the rays collect at a point on the retina, and a sharp image of the object is formed by the brain. But if the eye is not correctly focused, then the image forms either in front of the retina or behind it, depending on whether the rays are bent too much or too little, and the brain perceives an out-of-focus image of the object. In order to correct the focus, as mentioned in Chapter 2, "Fellow of the Royal Society", the eye naturally accommodates so as to be able to focus light rays on the retina from objects at different distances.

If Young was to grasp this mysterious process of accommodation, he needed to be able to monitor and measure it as accurately as possible. Over what range of distance from the eye could any particular eye accommodate and produce focused images? He therefore developed an optometer. Its principle was not new, having been suggested by Christoph Scheiner in 1619, nor was its implementation, which was carried out by Porterfield, but the accuracy and practicality of the instrument were greatly improved by Young.

Scheiner's observation was as follows (as translated into clearer language in the nineteenth century by Hermann Helmholtz). Make two pinholes in a card at a small distance apart less than the diameter of the pupil of the eye. Look through them with one eye as close to the pinholes as possible, keeping the other eye closed, at a small, clearly delineated object, such as a needle held in front of a bright window. Keep the object vertical, at right angles to the line joining the pinholes. Focus the eye on the object so that it is sharp. If you now shift the focus of your eye to something else, in front of the object or behind it, the object will appear double. Focus again on the object and the double images cross, coincide and appear single again.

Porterfield realized that this fact could be used to measure the *near point* of an eye—that is, the nearest to the eye an object can be brought while still remaining focusable and sharp (not double)—and the *far point*, the furthest distance away at which the object will remain sharp; and that these two points would give the range of the eye's accommodation. Instead of pinholes, Porterfield used narrow vertical slits to increase the visibility of the object, and a movable vertical slit in a lamp lit by a candle as the object to be viewed. When the eye saw the illuminated slit as a single line, rather than a double line, the eye was focused. Young's further development of the optometer was twofold: he incorporated a graduated scale giving the distance between the two slits (and the eye) and the single illuminated slit, and he added a convex lens close to the two slits. The purpose of this lens was to overcome the fact that the far point of a normal eye is at an infinite distance from the two slits, which clearly cannot be measured by the scale. The effect of the lens meant that all the distances measured by the optometer (not merely the far point), including the near point, had to be adjusted to give the 'true' focal distance—from which could be calculated the power of a spectacle lens required to correct short and long sight. Young made his optometers, which varied in size, out of both card and ivory; one still exists at the Royal Institution.

The near point of his own eye, after he had made the adjustment for the convex lens, turned out to be eight inches; and he took this to be normal. Today, a near point of ten inches is considered normal. This means that Young was somewhat short-sighted; the more short-sighted a person is, the nearer to the eye is his or her near point. In due course Young must have realized his myopia, because in his autobiographical sketch he writes: "He felt some inconvenience in society from being a little short sighted, and he used to attribute in part to this circumstance the mistakes which he sometimes made respecting the impression produced by what he said or did, on the feelings of others." (It seems possible that the frequent cases of mistaken identity in the dramatic plots of plays and operas of Young's age were more convincing to audiences then than they are now, because many people were short-sighted and did not wear spectacles.)

Young also experimented with the optometer to determine his far point. In the course of this process, he noted:

> My eye, in a state of relaxation, collects to a focus on the retina, those rays which diverge vertically from an object at the distance of ten inches from the cornea, and the rays which diverge horizontally from an object at seven inches distance. For, if I hold the plane of the optometer vertically, the images of the line appear to cross at ten inches; if horizontally, at seven. ... I have never experienced any inconvenience from this imperfection, nor did I ever discover it till I made these experiments; and I believe I can examine minute objects with as much accuracy as most of those whose eyes are differently formed.

Although Young did not name this condition (it was named three decades or so later by William Whewell, master of Trinity College, Cambridge, in a suggestion to the seriously astigmatic astronomer Sir George Biddell Airy), Young's comment is the first scientific recognition of astigmatism. Its name derives from the Greek for 'not at a point (*stigma*)'.

In an eye with astigmatism, the rays from a vertical line are focused differently to the rays from a horizontal line, and so the various rays do not collect at points in the same focal plane, with the result that the image is blurred. An optician tests for astigmatism by showing the patient a card with a series of radiating black lines. If astigmatism is present, one particular line will seem sharp, and the line at right angles to it will appear fuzzy.

When Young mentioned this experiment to a scientific instrument maker, William Cary, Cary told him that he had frequently observed the condition, and "that many persons were obliged to hold a concave glass obliquely, in order to see with distinctness, counterbalancing, by the inclination of the glass, the too great refractive power of the eye in the direction of that inclination". From this Young concluded that astigmatism was due to the crystalline lens in astigmatic eyes being at a slightly oblique angle to the vertical axis, and suggested that it could be compensated for by tilting a spectacle lens or the eyeglass of a telescope. While this is true, he was incorrect in dismissing the role of the cornea in astigmatism; today we know that corneal imperfections, a lack of symmetry in the curvature of the cornea so that different parts of it refract rays to slightly differing extents, are actually a much more common cause of astigmatism than misalignment of the crystalline lens.

Having found a relatively convenient way to measure the eye's focal distance with his optometer, Young now devised a method to measure the dimensions of his eye: its diameter and its length from back to front. His technique here, and in most of the experiments in his paper, was not for the clumsy or the faint-hearted and belongs in a long and honorable tradition of scientists experimenting on themselves. One must imagine Young, all alone in his house (still at this time in Norfolk Street) except perhaps for a servant, performing risky operations on his eyes, surrounded by candles, mirrors, lenses, microscopes, optometers and other homemade apparatus. "For measuring the diameters, I fix a small key on each point of a pair of compasses; and I can venture to bring the rings [of the keys] into immediate contact with the sclerotica [sclera]. The transverse diameter is externally 98 hundredths of an inch." To measure the distance from the back of the retina to the front of the cornea was rather more tricky (and painful!). He turned his eye inwards as far as it would go. Then he pushed the ring of one key in at the back of the eye and pressed on the back of the eyeball to produce the sensation of a bright spot on his retina—indicating that the key was almost touching the retina—in the center of his field of vision, coinciding with the direction of the eye's optical axis. "With an eye less prominent, this method might not have succeeded." Then, by looking into a mirror, he brought the second ring on the pair of compasses in contact with the cornea at the front of the eye. The distance between the back of the retina and the cornea turned out to be 91 hundredths of an inch, which was slightly less than the transverse diameter. Hence, the eyeball was not exactly spherical. From these figures, he could calculate the radius of curvature of the cornea. His measurements and calculations match extraordinarily well with modern measurements.

Then we come to the process of accommodation. Young wished to test four hypotheses for what happens during accommodation:

1. The curvature of the cornea changes.

2. The length of the eyeball changes.

3. Both changes occur at the same time.

4. The shape of the crystalline lens changes.

To test hypothesis one, he devised a series of experiments, of which we shall describe only two. The first consisted of a very careful examination with a graduated microscope of the reflection of a candle flame in the cornea of the eye of an assistant. Young's idea was to check whether the size of reflection varied as his assistant focused his eye on objects at different distances. It should have varied if the curvature of the cornea changed during accommodation, but stayed the same if the curvature remained the same. In his own words:

> I placed two candles so as to exhibit images in a vertical position in the eye of Mr König, who had the goodness to assist me; and, having brought them into the field of the microscope, where they occupied 35 of the small divisions, I desired him to fix his eye on objects at different distances in the same direction: but I could not perceive the least variation in the distance of the images.

The second, and the most crucial of all the experiments, involved immersing the eye in water. As underwater swimmers are aware, the unaided eye cannot focus sharply in water. The explanation for this is that light passing through the water-cornea interface into the aqueous humor is no longer refracted because the aqueous humor, optically speaking, is very nearly equivalent to water. (Recall that it is the air-cornea interface that causes about two-thirds of the refraction in the eye.) Young reasoned that if, in front of an eye in water, he were to add a lens with a refractive power equivalent to the eliminated air-cornea interface, the eye should be able to focus again. If, in addition, the immersed eye with the extra lens could still accommodate, then the process of accommodation could not involve the cornea.

Here is how he describes the experiment:

> I take out of a small botanical microscope, a double convex lens, of eight-tenths radius and focal distance, fixed in a socket one-fifth of an inch in depth; securing its edges with wax, I drop into it a little water, nearly cold, till it is three fourths full, and then apply it to my eye, so that the cornea enters halfway into the socket, and is everywhere in contact with the water. My eye immediately becomes presbyopic [i.e., long-sighted, because the loss in refraction at the cornea means that the image forms behind the retina], and the refractive power of the lens, which is reduced by the water ... is not sufficient to supply the place of the cornea, rendered inefficacious by the interventions of the water; but the addition of another lens of five inches and a half focus, restores my eye to its natural state and somewhat more. I then apply the optometer, and I find the same inequality in the horizontal and vertical refractions as without the water [demonstrating that Young's astigmatism was not corneal in origin]; and I have, in both directions, a power of accommodation equivalent to a focal length of four inches, as before ... After this it is almost necessary to apologize for having stated the former experiments; but, in so delicate a subject, we cannot have too great a variety of concurring evidence.

With his first hypothesis now abandoned, Young designed experiments to test hypothesis two: that the eyeball changed its length during accommodation, like a camera lens adjusting for focus. His method was to fix the length of the eyeball mechanically, so that it could not expand or contract, and then try to focus his eye on objects at different distances. If, under these conditions, his eye could still accommodate, then accommodation could not be due to the change in length of the eyeball.

The most important of this second group of experiments was described by Young, somewhat unnervingly, as follows:

> [A] much more delicate [test], was the application of the ring of a key at the external angle, when the eye was turned as much inwards as possible, and confined at the same time by a strong oval iron ring, pressed against it at the internal angle. The key was forced in as far as the sensibility of the integuments would admit, and was wedged, by a moderate pressure, between the eye and the bone. In this situation the phantom [the bright spot on the retina] caused by the pressure extended within the field of perfect vision, and was very accurately defined.

With the eye held in this state, he argued, the phantom was a highly sensitive indicator of the length of the eyeball. The slightest increase or decrease in length would increase or decrease the pressure on the retina, and alter the size and shape of the phantom. (His paper includes his drawings of this illusion under various conditions.) "But no such circumstance took place; the power of accommodation was as extensive as ever; and there was no perceptible change either in the size or in the figure of the oval spot."

Hypothesis three—that accommodation was due to a mixture of change in corneal curvature and eyeball length—was now obviously ruled out. By process of elimination, it appeared that hypothesis four—change in the shape of the crystalline lens—was the most likely to be true. Young set about trying to find experimental evidence for it.

Home, with the help of Ramsden, had claimed in 1795 that a man they had examined named Benjamin Clerk, whose eye had been couched because he had a cataract, afterwards retained the power of accommodation. Clerk, a seafarer, had gone missing and was unavailable to Young in 1800, but an optician friend of Young's, a Mr Ware, introduced him to five of his patients who he thought might make suitable subjects. Ware had originally been convinced by Home's paper but subsequently had noticed that all of his patients who had been couched derived "obvious advantage" from using two kinds of spectacles—one for close-up work such as reading, the other for seeing at a distance. This strongly implied they had a deficiency in their power of accommodation. (Bifocal spectacles were invented to deal with this problem in older people, whose eyes generally lose some power of accommodation in their forties.)

None of the five subjects was perfect for Young's purposes, as he made clear in his report on each of them. But after very carefully testing them all with his optometers, in the presence of their optician Mr Ware, he came to a firm conclusion: "the universal result is, contrary to the expectation with which I entered on the inquiry, that in an eye deprived of the crystalline lens, the actual focal distance is totally unchangeable." This would be fully confirmed only when Benjamin Clerk was located again some years later and tested in the presence of Home, Young and two others. Their joint examination was extremely painstaking—not least because the reputations of the physicians as scientists were at stake—and Young noted with satisfaction in his *Introduction to Medical Literature*, published in 1813, that "the imperfect eye, from which the crystalline lens had been extracted, possessed no power whatever of altering its focus, while the same tests exhibited a very considerable change in the focal distance of the perfect eye".

In his pioneering 1800 paper, he had thus established, at least provisionally, the process of accommodation of the eye; but he still needed to put forward a mechanism for the process. The crystalline lens must change shape, he had now shown, but how did it do this? In 1793, he had maintained that the lens was muscular. In 1800, he was less certain, after re-examining the anatomy of the eye: "Now, whether we call the lens a muscle or not, it seems demonstrable, that such a change of figure [shape] takes place as can be produced by no external cause; and we may at least illustrate it by a comparison with the usual action of muscular fibers." Here Young was less perceptive, and it would be left to others, including Helmholtz, later in the century, to identify the thin threads of robust material, called zonules, that hold the non-muscular crystalline lens in place and are attached to the ciliary body housing the ciliary muscle.

However, Young was careful *not* to include the muscular lens hypothesis in his final paragraph summing up what he saw as the definite results achieved in his paper "On the mechanism of the eye". This paragraph has often been quoted by physiologists for its clarity and concision:

First, the determination of the refractive power of a variable medium, and its application to the constitution of the crystalline lens. Secondly, the construction of an instrument for ascertaining, upon inspection, the exact focal distance of every eye, and the remedy for its imperfections. Thirdly, to show the accurate adjustment of every part of the eye, for seeing with distinctness the greatest possible extent of objects at the same instant. Fourthly, to measure the collective dispersion of colored rays in the eye. Fifthly, by immerging [immersing] the eye in water, to demonstrate that its accommodation does not depend on any change in the curvature of the cornea. Sixthly, by confining the eye at the extremities of its axis, to prove that no material alteration of its length can take place. Seventhly, to examine what inference can be drawn from the experiments hitherto made on persons deprived of the lens; to pursue the inquiry on the principles suggested by Dr Porterfield; and to confirm his opinion of the utter inability of such persons to change the refractive state of the organ. Eighthly, to deduce, from the aberration of the lateral rays [astigmatism], a decisive argument in favor of a change in the figure of the crystalline; to ascertain, from the quantity of this aberration, the form into which the lens appears to be thrown in my own eye, and the mode by which the change must be produced in that of every other person.

Young's other major contribution to understanding the eye came in a second lecture, "On the theory of light and colors", given to the Royal Society almost exactly a year later (after he had moved from Norfolk to Welbeck Street), and published in 1802. This is where he put forward his theory of three-color vision. But its presentation could hardly be more different from the detailed experimentation and calculation documented in the first lecture. His far-sighted idea was more like an intuition, an *aperçu*, than a developed theory—and it would take a century and a half before it was verified experimentally. "Surely the most prescient work in all of psychophysics", a physicist called it in 1989 (Walter Moore, in his scientific biography of Erwin Schrödinger, another physicist with wide interests, including color vision). Young himself thought so comparatively little of the idea that he did not even mention his three-color theory in his list of publications at the end of his autobiographical sketch.

In the seventeenth century, Newton had split white light into the colors of a spectrum with a prism and reconstituted the spectrum into white light with a second prism; he had also used a second prism to show that the individual colors of the spectrum could not be further split. In 1672, Newton introduced the term 'primary colors', and pondered how many such colors there were, favoring seven, and how discrete colors might relate to a clearly continuous spectrum, in his *Opticks*, published in 1704. During the eighteenth century, the concept of primary colors became generally accepted, but they were reduced to three in number, usually red, yellow and blue. Yet there was no understanding of how these primary colors could create the great variety of hues—more than 150 of them—distinguishable by the eye, mainly because the relationship between color and wavelength was not appreciated, in the absence of acceptance of a wave theory of light.

Young's brilliant insight, stimulated by his embrace of the undulatory/wave theory during 1801, was to imagine how the retina might actually detect the sensation of color. He wrote:

> Now, as it is almost impossible to conceive each sensitive point of the retina to contain an infinite number of particles, each capable of vibrating in perfect unison with every possible undulation, it becomes necessary to suppose the number limited, for instance, to the three principal colors, red, yellow, and blue, of which the undulations are related in magnitude nearly as the numbers 8, 7, and 6; and that each of the particles is capable of being put in motion less or more forcibly by undulations differing less or more from a perfect unison; for instance, the undulations of green light being nearly in the ratio of 6.5, will affect equally the particles in unison with yellow and blue, and produce the same effect as a light composed of those two species; and each sensitive filament of the nerve may consist of three portions, one for each principal color.

In other words, the brain would perceive red light, with the longest wavelength, as red because it would stimulate (be in "perfect unison" with) only one type of receptor ("particle") in the retina; ditto for yellow light, with a shorter wavelength, which would stimulate only a second type of receptor; and for blue light, of an even shorter wavelength, which would stimulate only a third type of receptor. Light of an intermediate wavelength, such as green, would stimulate both the yellow and the blue receptors, though less strongly than yellow and blue light; and the mixture of the two sensations would be perceived as green in the brain. Young had "proposed a theory of color vision which, for the first time, suggested the brain may not only receive information, but [that] it processes and integrates the information it receives", the physiologist J. Z. Young observed.

The following year, 1802, as a result of experiments on the color spectrum by the physicist and chemist William Hyde Wollaston, Young changed his choice of "principal colors" to which the retina was sympathetic from red, yellow and blue to red, green and violet. This is the work that lay fallow in the *Philosophical Transactions* of the Royal Society until it was rediscovered by an excited Helmholtz in the 1850s and developed into the Young-Helmholtz theory of color vision, which was soon confirmed and extended by the experiments of James Clerk Maxwell with spinning tops painted with sections of different color (an idea which Young had also written about). Yet it took until 1959 before scientists made "the definitive experiments that finally proved Young's idea that color must depend on a retinal mosaic of three kinds of detectors," comments David Hubel, one of today's authorities on visual neuroscience. The experiments were the work of two groups in the United States—those of George Wald and Paul Brown at Harvard University, and of William Marks, William Dobelle and Edward MacNichol at Johns Hopkins University—who examined the cones in the retina and their ability to absorb light of different wavelengths and discovered just three cone types, as speculated by Young in 1801.

Wald went on to explain color blindness in terms of a reduced or absent receptivity in one or more of the three cone types. Here again Young had led the way. He was interested in the color blindness of one of his contemporaries, the chemist John Dalton, who in 1798 stirred great interest by describing how red, orange, yellow and green were akin to him, but how he could distinguish blue and purple. Dalton was convinced that the cause was his vitreous humor being tinged blue (and therefore absorbing red light before it reached the retina). But Young did not agree with Dalton's notion, remarking in his published Royal Institution lectures that "this has never been observed by anatomists, and it is much simpler to suppose the absence or paralysis of those fibers of the retina which are calculated to perceive red; this supposition explains all the phenomena". Dalton's vitreous humor was tested after his death in 1844 (at his written request) and found to be colorless, supporting Young; but when the retina from one of his preserved eyes was examined in the 1990s, the evidence was less supportive of Young, since it lacked the photo-pigment sensitive to light of middle wavelength rather than the longer-wavelength red light.

The moment has now arrived to leave Young's contributions to physiology and turn to his lectures on physics and related subjects at the Royal Institution in 1802-03. The polymath was about to face the public in London for the first time. The encounter would prove to be a disturbing one, both for Young and for his listeners.

Chapter 6

Royal Institution Lecturer

"I shall esteem it better to seek for substantial utility than temporary amusement; for if we fail of being useful, for want of being sufficiently popular, we remain at least respectable: but if we are unsuccessful in our attempts to amuse, we immediately appear trifling and contemptible."

Young, introduction to A Course of Lectures on Natural
Philosophy and the Mechanical Arts, *1807*

Great thinkers do not always make great lecturers. A Nobel Prize is no guarantee of its possessor's ability to communicate complex ideas. Among physicists, for example, Albert Einstein was a clear and witty speaker, who relished tough questions from the audience, whereas Niels Bohr, Einstein's close intellectual rival, was notoriously hard to follow, with a tendency to give baffling answers to simple questions. In the 1770s, John Hunter's lectures "split London's surgical fraternity down the middle: one half acclaimed Hunter as a genius and his lectures as inspired, the other half condemned him as a charlatan and his lectures as incomprehensible," according to Hunter's biographer Wendy Moore. It was said at the time that one of Hunter's lectures was so scantily attended that he was obliged to have a skeleton brought in so that he could begin the lecture in the expected way, "Gentlemen...". One can only too easily imagine how much Hunter's hostile surgeon colleagues must have enjoyed spreading that particular story around London.

Moreover, great minds who lecture well to their peers do not necessarily carry conviction with a general audience. Think of the fellow philosophers Bertrand Russell and Ludwig Wittgenstein. Russell was one of the most celebrated public speakers of his age, whereas Wittgenstein was famous only within certain subsets of professional philosophy. Even Einstein was not a natural popularizer of his difficult ideas; the theory of relativity had to be explained to the world (and even to physicists) mainly by others with a gift for communication, such as Sir Arthur Eddington.

Young's experience as a lecturer was reminiscent of Hunter's. Young's lectures divided the fellows of the Royal Society and other scientists of his day into admirers willing to struggle through the thicket of his presentation, and detractors inclined to dismiss his incomprehensibility as something close to charlatanism. As for the ability to popularize, Young must rank low among scientists, especially as compared with his scintillating contemporaries Sir Humphry Davy and Michael Faraday, who established a unique tradition of lecturing on science to all and sundry at the Royal Institution using exciting and instructive demonstrations.

As Young himself admitted to his audience at the end of his first course of lectures in May 1802, while paying generous tribute to Davy:

> [M]y colleague ... even in his first course, has been able to unite in an unprecedented degree perspicuity of theory with brilliancy of experimental illustration. I will not enlarge on what I wish my own lectures to be lest I should hereafter fall short of my professional intentions: but I must at least beg you to consider yourselves as having been admitted into the study of a painter, while he is tracing the outlines on his canvas, and laying on the first masses of coarse coloring, in a state in which no artist would without reluctance exhibit his productions even to the best judges.

By the time Young wrote his autobiographical sketch a quarter of a century later, he would refer to the language he used in his Royal Institution lectures as being "never either very popular or very fluent", and "his compressed and laconic style and manner" as being "more adapted for the study of a man of science than for the amusement of a lady of fashion."

A director of the Royal Institution in the late twentieth century took such self-criticism to mean that Young was "a narcoleptically boring speaker", and Davy a "coruscatingly brilliant" one. But this seems too harsh on Young, given the rather limited surviving evidence. And even if Young really did lull his audiences to sleep with his delivery, he had the satisfaction of knowing that his content was original and far reaching. A history of the Royal Institution written in 1871, post-Faraday, noted that Young's lectures "must even now be held to rank as the greatest work in the literature of the institution." Their stock has maintained its value since the late nineteenth century, in contrast to Davy's discourses, which were of little interest to scientists by the twentieth century.

The Royal Institution was founded in 1799 by Count Rumford, a physicist of importance and also a determined, flamboyant and unscrupulous American of royalist sympathies, originally born plain Benjamin Thompson into a farming family in Massachusetts, who had obtained his title as a count of the Holy Roman Empire while acting as war and police minister for the elector of Bavaria. Unlike the Royal Society, with its emphasis on pure science as in Newton's *Principia*, the Royal Institution was established in order to study and promote the application of science to society. Rumford's original prospectus of March 1799 described itself as "Proposals for forming by subscription, in the metropolis of the British Empire, a public institution for diffusing the knowledge, and facilitating the general introduction, of useful mechanical inventions and improvements; and for teaching, by courses of philosophical lectures and experiments, the application of science to the common purposes of life." Fifty-eight men of influence quickly subscribed fifty guineas each, including one duke, six earls, seven lords, eleven knights, one bishop and eighteen members of Parliament. Although physics (natural philosophy) and chemistry were very much on the Royal Institution's initial agenda for study, so too were bread-making, the production of cheap and nutritious soups for feeding the poor, the design of cottages and cottage fireplaces and of kitchen fireplaces and kitchen utensils, and numerous other practical aspects of the "common purposes of life".

The president of the Royal Society, Sir Joseph Banks, was most supportive of the new institution. Young was a regular attendee at Royal Society meetings, where he and Banks became friendly. It was Banks who recommended Young to Rumford to be professor of natural philosophy in 1801, after the resignation of the first professor, Thomas Garnett, who was ill and embittered with Rumford. Davy, in the meantime, had already been appointed to lecture in chemistry and made a good impression despite his uncouth background; he was promoted to professor of chemistry in 1802.

Young and Rumford were of one mind on Rumford's controversial theory of heat (we shall leave this to Chapter 8), and Young was excited by Rumford's offer, though immediately concerned about its effect on his fledgling medical career, as we already know from his letter to Dalzel just after moving into his new house in June 1801. But with his inheritance from his great-uncle in hand—£10,000 in 1797—Young could afford to bargain a little. He felt that, though untried, he was worth the same salary as Garnett. In early July, Young wrote to Rumford from Welbeck Street:

> I am willing to undertake the various charges which you have the goodness to detail, and I flatter myself that you will have no reason to complain of any want of zeal on my part in the service of the Royal Institution. ... But I confess I think it would be in some measure degrading both to me and to the institution that the salary which appears to me to have been no more than moderate before, should now be reduced by one-fourth, at the same time that the labor and responsibility of the employment are rather increased than lessened.

This referred to the fact that as well as lecturing on natural philosophy, Young was expected to edit the house journal and also act as general superintendent of the house. After this, he added a paragraph about his medical work, in the slightly stiff and circumlocutory language that he sometimes fell into:

> It would not be my wish, and the duties of the professorship would certainly render it impossible for me to attempt any extent of medical practice; but I should be sorry to bind myself to reject the little that might accidentally fall in my way. I do not mention this as a matter of any consequence, but to avoid having it understood, from the conversation I had with you, that I should be obliged to refuse my advice to a friend who might consult me.

Young got his way, the salary was maintained, and it was soon agreed by the managers of the Royal Institution, at the suggestion of Rumford, that Young should be hired at £300 per annum.

This turned out to be a very good deal for the managers. For the next nine months and more, Young worked fanatically hard on his lectures, and gave himself totally to the work of the Royal Institution. He was not a man prone to exaggeration, as must be transparent to the reader by now, yet he told Dalzel in late March 1802 that "an immediate repetition of the labor and anxiety that I have undergone for the last twelve months would at least make me an invalid for life."

During the first half of that year, between 20 January and 17 May, on Mondays and Wednesdays at 2 p.m. and on Fridays at 8 p.m., Young gave 50 lectures on different subjects. He then repeated these lectures in the first half of 1803 and probably added more, because in the final publication of the lectures in 1807, there are 60 lectures specified. They were classified into the following parts—"Mechanics", "Hydrodynamics", "Physics" and "Mathematics"—but their scope was much wider than these titles suggest, as noted by Nicholas Wade in his introduction to the 2002 reprint: "For example, the first includes drawing and architecture, the second music and optics, the third astronomy and geography, the fourth pure and applied mathematics."

It is less taxing for the reader if we treat the scientific content of the lectures separately from their social context, so let us leave the science till Chapter 8, which deals purely with the publication of the lectures. Here we shall consider only Young's "Introduction", in which he defined his aims and his target audience.

He first pays tribute to the "primary and peculiar object of the Royal Institution", which is "to apply to domestic convenience the improvements which have been made in science, and to introduce into general practice such mechanical inventions as are of decided utility." But it is quickly apparent that he is speaking more out of lip service than conviction. For he continues:

> To exclude all knowledge but that which has already been applied to immediate utility, would be to reduce our faculties to a state of servitude, and to frustrate the very purposes which we are laboring to accomplish. No discovery, however remote in its nature from the subjects of daily observation, can with reason be declared wholly inapplicable to the benefit of mankind.

In modern parlance, pure scientific research has an unpredictable way of becoming applied science—for instance, the apparently 'useless' laser invented by physicists in the late 1950s from a half-developed theory of Einstein. Young was unequivocally on the side of science for science's sake:

> Those who possess the genuine spirit of scientific investigation, and who have tasted the pure satisfaction arising from an advancement in intellectual acquirements, are contented to proceed in their researches, without inquiring at every step what they gain by their newly discovered lights, and to what practical purposes they are applicable: they receive a sufficient gratification from the enlargement of their views of the constitution of the universe, and experience, in the immediate pursuit of knowledge, that pleasure which others wish to obtain more circuitously by its means. And it is one of the principal advantages of a liberal education, that it creates a susceptibility of an enjoyment so elegant and so rational.

It is perhaps hardly necessary to say that Young was quintessentially a figure of the Enlightenment.

As for his lecturing style:

> I shall in general entreat my audience to pardon the formality of a written discourse, in favor of the advantage of a superior degree of order and perspicuity. ... The most difficult thing for a teacher is, to recollect how much it cost himself to learn, and to accommodate his instruction to the apprehension of the uninformed: by bearing in mind this observation, I hope to be able to render my lectures more and more intelligible and familiar; not by passing over difficulties, but by endeavoring to facilitate the task of overcoming them; and if at any time I appear to have failed in this attempt, I shall think myself honored by any subsequent inquiries that my audience may be disposed to make.

In other words, Young was happy to take questions from his audience—though there is no available record of how well he answered them.

The listeners he had in mind included ladies, as he made clear:

> A considerable portion of my audience, to whose information it will be my particular ambition to accommodate my lectures, consists of that sex which, by the custom of civilized society, is in some measure

exempted from the more laborious duties that occupy the time and attention of the other sex. The many leisure hours which are at the command of females in the superior orders of society may surely be appropriated, with greater satisfaction, to the improvement of the mind and to the acquisition of knowledge, than to such amusements as are only designed for facilitating the insipid consumption of superfluous time. … In this point of view the Royal Institution may in some degree supply the place of a subordinate university, to those whose sex or situation in life has denied them the advantage of an academical education in the national seminaries of learning.

Plainly, there was more than enough room in this introduction for misunderstanding and mismatch between lecturer and audience. Young was setting out to do the impossible, as any television producer of the annual Royal Institution lectures two centuries later could have told him. The not-very-friendly tutor at Emmanuel College Cambridge noted: "I remember … his taking me with him to the Royal Institution, to hear him lecture to a number of silly women and dilettanti philosophers. But nothing could show less judgment than the method he adopted; for he presumed, like many other lecturers and preachers, on the knowledge and not on the ignorance of his hearers."

A wicked caricature of a lecture at the Royal Institution drawn by James Gillray, published on 23 May 1802 and reproduced here (see Figure 6.1), catches an atmosphere of earnestness mixed with farce. Facetiously entitled "Scientific Researches!—New Discoveries in Pneumaticks!—or—an Experimental Lecture on the Powers of Air", it shows one lecturer determinedly administering gas to the mouth of a guinea pig, Sir J. C. Hippisley, one of the managers of the institution (and a well-known snob), and producing a most disastrous effect in his breeches. The other lecturer, chubby cheeked, with a mischievous glint in his eyes, holds a powerful pair of smoking bellows. In the audience, Count Rumford, standing at the far right, watches enigmatically, while various male members of the upper classes attempt to maintain a scholarly solemnity, a 'silly woman' throws up her hands theatrically, and some members of the lower orders, male and female, treat this serious scientific affair more like a music-hall turn.

Figure 6.1   *Young (with Humphry Davy) as a lecturer at the Royal Institution, carica-
tured by James Gillray in 1802. Rumford stands at far right.*

The first lecturer is likely to be Young, but could be Garnett; the second is unquestionably Davy. Since Gillray did not identify anyone by
name, we cannot be sure. The face looks quite like that of the only known
portrait of Young, and the date, 1802, is compatible with Young, not
Garnett. On the other hand, we know from the diary of Lady Holland that
in March 1800, Garnett administered the recently discovered laughing gas
(nitrous oxide) to Hippisley and that "the *effect* on him was so *animating*
that the ladies tittered, held up their hands, and declared themselves satisfied." But this date is not compatible with the presence in the caricature of
Davy, who did not start at the Royal Institution until 1801. Probably, therefore, Gillray did not depict a particular lecture but used artistic license to
combine details from several reported incidents.

What is patent is that the expectation of Royal Institution audiences
tended more to entertainment than to expositions filled with Young's pro-
fessed "elegance" and "reason". And this seems to have applied to the managers, too, who, after the departure of a disillusioned Rumford to France in

1802, aimed to draw a fashionable crowd. (In 1804, to the disgust of the president of the Royal Society, Banks, the Royal Institution hosted lectures by the poet Samuel Taylor Coleridge, the painter John Landseer and the future canon of St Paul's Cathedral, Sydney Smith, who spoke on moral philosophy to large audiences.) It had to be merely a matter of time before Young himself would depart; and the break finally occurred in the summer of 1803. He gave his reason for resigning as the conflict between his duties at the Royal Institution and the needs of his medical practice, which was perfectly consistent with the concerns he expressed when taking the job in 1801. It is, however, probable that the managers wanted him out, given the offhand way in which they treated him during the period before he resigned. Nevertheless the break was amicable: Young accepted life membership of the Royal Institution and continued to publish in its journal to the end of his life.

His medical practice had not been utterly neglected in this period. During the summer of 1802, after he finished lecturing in May, Young agreed to accompany the two great-nephews of his friend the duke of Richmond to France, which was briefly at peace with Britain. His company was wanted partly to keep an eye on their health and partly because he spoke French and their tutor did not. During the three months they stayed in Rouen, Young spent two weeks visiting Paris, and attended the scientific discussions of the National Institute. Napoleon Bonaparte himself was there. Napoleon had a genuine interest in science. "If I had not had to conquer the world," he is apocryphally supposed to have remarked to the mathematician Lagrange, "I should have become a scientist and discovered it." ("Sire, Newton has already done that," Lagrange supposedly replied, "and there is only one world to discover.") According to his biographer Peacock, Young was actually introduced to Napoleon, but Young himself says only that he had the "amusement of hearing Napoleon take part in the discussions". Anyway, the visit was excellent preparation for his role as foreign secretary of the Royal Society, which he became soon after resigning from the Royal Institution, in early 1804, and for his later long and fruitful association with French physics.

If Young was disappointed about his parting with the Royal Institution, he did not show it. In the second half of 1803, rather than going straight back into medicine, he threw himself into work on his next

Royal Society lecture, entitled "Physical optics". It would prove to be his most important work ever, his *pièce de résistance*, including as it did his "experimental demonstration of the general law of the interference of light". But to understand how he came to this, we must first go back a few years.

Chapter 7

Let There Be Light Waves

*"The theory of light and colors, though it did not occupy a large por-
tion of time, I conceive to be of more importance than all that I have
ever done, or ever shall do besides."*

Young, letter to Andrew Dalzel, 1802

All natural philosophers in the eighteenth century fascinated by the
mystery of light worked in the long shadow of Sir Isaac Newton,
whether they liked it or not. Newton's analysis of mechanics and
gravity, published in his *Principia* in the 1680s, had revolutionized natural
philosophy, and remains the basis of physics in the twenty-first century.
Given that book's heroic and unchallenged prestige by the time of
Newton's death, his *Opticks*, published in 1704, though far more speculative than the *Principia*, was destined to dominate all scientific discussion of
light. Young, having first read both books in 1790, when he was 17, was
steeped in them by the time he started his own investigations of sound and
light in Cambridge in 1797. Not only did he personally revere Newton's
work, but he also knew that the general hero worship of Newton meant he
was certain to be severely criticized if he were to question Newton's authority, especially in a matter as fundamental as the very constitution of light.

"I have ... been accused of insinuating 'that Sir Isaac Newton was but
a sorry philosopher.' But it is impossible that an impartial person should
read my essays on the subject of light without being sensible that I have as
high a respect for his unparalleled talents and acquirements as the blindest
of his followers, and the most parasitical of his defenders," Young felt
obliged to state in 1804 in reply to a vitriolic critic of his theory of the

interference of light. "But, much as I venerate the name of Newton, I am not therefore obliged to believe that he was infallible. I see, not with exultation, but with regret, that he was liable to err, and that his authority has, perhaps, sometimes even retarded the progress of science."

Theories of light as being either a wave or a particle go back to the ancient Greeks. Newton, of course, favored a particle theory of light—though he was not completely convinced that this was correct, as we shall shortly see. The only major proponent of a wave theory in Newton's time was the Dutch physicist and astronomer Christiaan Huygens, who published his ideas in 1678. In Newton's corpuscular theory, light was imagined to be a stream of minute particles or 'corpuscles' emitted by a light source, shooting through empty space like bullets in the form of light rays, and detected by their impact on the retina of the eye. In Huygens's undulatory theory, light was thought to be a wave spreading out in all directions from a light source, transmitted via a medium known as the *ether*, and detected by the wave's creation of sympathetic vibrations in the retina. Light waves were regarded as analogous to the sound waves that spread from a struck tuning fork, with the retina taking the place of the tympanum as detector. In the case of sound, the medium that undulated was known to be the air, whereas with light it was supposed to be the ether.

The ether permeated the entire universe, including all matter. For various respectable physical reasons, it had to be "absolutely stationary, weightless, invisible, with zero viscosity, yet stronger than steel and undetectable by any instrument", in the words of a current theoretical physicist, Michio Kaku. Einstein, whose relativity theory finally killed off the ether as a scientific concept in the years after 1905, was not surprised that Newton distrusted it and adhered to a theory of light in which the role of the ether was not essential:

> The assumption that space was filled with a medium consisting of material points that propagated light waves without exhibiting any other mechanical properties must have seemed to him quite artificial. The strongest empirical arguments for the wave theory of light— fixed speeds of propagation, interference, diffraction, polarization— were either unknown or else not known in any well-ordered synthesis. He was justified in sticking to his corpuscular theory of light.

Let us see how the two radically different theories dealt with the physical phenomena of light known to Newton and Young in the eighteenth century, before we come to Young's own discoveries. Which theory explained best the propagation of light and its various kinds of 'bending': reflection, refraction and diffraction? (Polarization, which Einstein mentions, will be left until Chapter 11, "Waves of Enlightenment", where it is crucial in understanding what kind of a wave light is.)

The simplest phenomenon requiring explanation was that light is transmitted through a medium—such as air or interplanetary space—in straight lines. This is an everyday observation, visible in the shafts of sunlight coming through a cloud, in the formation of sharp-edged, deep-black shadows, and in the fact that a lighted lantern or candle vanishes when some object obstructs the shortest path between the light and the eye of the observer. Rectilinear propagation was also confirmed by astronomical events, for example solar eclipses. Such behavior was self-evident for a stream of corpuscles but not for a wave. Water waves that rippled from a stone dropped into a pond could be seen to spread in all directions and to bend around obstacles to some extent; and sound waves could be heard to bend, otherwise how could two people invisible to each other manage to converse through a half-closed door? Light, however, did not appear to bend. In Newton's emphatic words in his *Opticks*: "sounds are propagated as readily through crooked pipes as through straight ones. But light is never known to follow crooked passages nor to bend into the shadow. For the fixed stars by the interposition of any of the planets cease to be seen. And so do the parts of the sun by the interposition of the moon, Mercury or Venus." Elsewhere, he wrote: "To me the fundamental supposition itself seems impossible, namely, that the waves or vibrations of any fluid can, like rays of light, be propagated in straight lines without a continual and very extravagant spreading and bending every way into the quiescent medium where they are terminated by it."

So much for the propagation of light, which appeared to favor the corpuscular theory. Moving on to reflection, it was necessary for the two theories to explain the well-known fact that when a light ray strikes a flat reflecting surface like a mirror, the angle of incidence is equal to the angle of reflection, as shown in Figure 7.1. In the corpuscular theory, the explanation was straightforward: the corpuscles would behave like billiard balls bouncing off the cushion of a billiard table at equal angles. The wave theory, too, had no real difficulty; once Huygens assumed that a light ray could be mathematically modeled as the path of a point on the wave front, he could easily deduce the law of reflection. Neither theory was favored in explaining simple reflection.

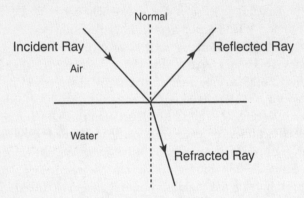

Figure 7.1   *Law of reflection and refraction of light.*

Refraction was a more decisive test. When light rays strike the surface of water and pass through it, the angle of incidence and the angle of refraction differ, as shown too in the figure. The angle of refraction is less than the angle of incidence, and the light ray is bent toward the normal (the perpendicular line). Although the relationship between the two angles was formulated in Snell's law as early as 1621 (by the mathematician Willebrord Snell), in Newton's day it still needed a physical explanation. Using Snell's law, one could now calculate the angle of refraction if one knew a light ray's angle of incidence. But what actually caused the light ray to bend, and why did it bend toward the normal?

Newton's answer was not very convincing. He proposed that the velocity of light in water was faster than in air—an idea opposed to common-sense expectation, given that water is a denser medium than air, and one would naturally expect a denser medium to slow the velocity of corpuscles rather than speed them up. The moment the corpuscles entered the water they were acted upon by a force, said Newton, that pulled them toward the normal, increasing their velocity and altering their direction of motion.

But the nature of this force was inexplicable, and unlike the gravitational force, there was no supporting evidence for it from phenomena other than refraction. Huygens, on the other hand, assumed the opposite to Newton: that light travels slower in water than in air. He was then able to use the wave theory in a direct and simple way, without the need to postulate any new force, to calculate Snell's law.

So the wave theory was definitely favored in explaining refraction. However the verdict was by no means conclusive. The evidence that would have clinched it was not yet available. What precisely was the velocity of light in air and in water? The first modern estimate of the velocity of light was made in the 1670s by astronomical measurements, but not until 1850 were experiments (by Armand Fizeau and Léon Foucault) able to measure it accurately and to prove that light moves more slowly in water than in air, as Huygens had assumed. This measurement was a sort of capstone to the edifice of the wave theory of light built on Young's initial experiments during the first half of the nineteenth century.

When light falls on water, a proportion is refracted and the rest is reflected, as is obvious from the shimmering surface of a swimming pool in bright sunshine. With the wave theory, this split is easy to explain, as it is common to all kinds of wave—for instance, sound waves, which are both transmitted through a wall and reflected by it in the form of a reverberation or echo. Why, though, should one corpuscle pass into the water and be refracted, while another identical corpuscle should instead be reflected? There seemed to be no good physical reason. The conundrum was embarrassing for Newton and forced him to make a frankly contrived suggestion that the first corpuscle was in a "fit of easy transmission" and the second was in a "fit of easy reflection". Moreover, to confuse the picture further, Newton attributed these corpuscular "fits" to the existence of waves! It is worth quoting his actual words here, from the last part of his *Opticks*, query 17, where he draws an analogy between water, sound and light:

> If a stone be thrown into stagnating water, the waves excited thereby continue some time to arise in the place where the stone fell into the water, and are propagated from thence in concentric circles upon the surface of the water to great distances. And the vibrations or tremors excited in the air by percussion, continue a little time to move from the place of percussion in concentric spheres to great distances. And

in like manner, when a ray of light falls upon the surface of any pellu-
cid body, and is there refracted or reflected, may not waves of vibrations, or tremors, be thereby excited in the refracting or reflecting
medium at the point of incidence... and are not these vibrations
propagated from the point of incidence to great distances? And do
they not overtake the rays of light, and by overtaking them successively, do they not put them into the fits of easy reflection and easy
transmission described above?

It is plain from this that Newton himself realized the corpuscular theory
alone could not explain all the phenomena of light.

The strange 'fits' were used by Newton to explain an important set of
extraordinarily detailed observations in the second part of his *Opticks*,
which quickly became known as Newton's rings. Today, the rings are used
for quality-testing the uniformity of a polished lens surface by bringing it
in contact with a perfectly flat glass surface. Newton was led to see the rings
by the beautiful colors visible on soap bubbles or when an oil film on a
rainwater puddle catches the light. He was intrigued that although the
soapy liquid was naturally colorless, when the soap film became very thin
in a bubble it could be brilliantly colored. Indeed, if a soap film is blown
covering the mouth of a wine glass and then the glass is fixed on its side so
that the film is now vertical, as the soap solution gradually drains away,
and the film becomes thinner and thinner, bands of color are seen to shift
on its surface, and the thinnest part at the top goes completely black just
before it bursts. Newton's experiment was far more controlled and exact
than this, involving a thin film of air (not a soap film) trapped between a
perfectly flat glass plate and a very slightly convex lens, and illuminated
from above the lens by a vertical beam of white light. The light rays were
reflected from two surfaces, that of the lens-air interface at the top of the
air film and that of the air-glass plate interface at the bottom of the air
film. The reflected light, when viewed with the eye over the center of the
lens, produced a set of concentric colored rings centered on the point of
contact between the lens and the glass plate. Viewed from the other side
(that of the glass plate), the transmitted light also produced a set of concentric colored rings, but with the colors complementary to the colors in
the first set of rings. Figure 7.2 shows the colors, as given by Newton.
He explained the pattern at elaborate length with the following awkward

corpuscular concept: "The returns of the disposition of any ray to be reflected I will call its fits of easy reflection, and those of its disposition to be transmitted its fits of easy transmission, and the space it passes between every return and the next return, the interval of its fits." We shall later see how Young reinterpreted Newton's data in terms of the interference of light and thereby explained Newton's rings more simply with the wave theory.

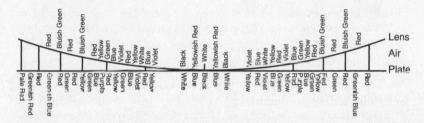

Figure 7.2   *Colors in Newton's rings, visible when a glass convex lens is placed in contact with a glass plate. A thin film of air is trapped in between.*

Finally, having considered reflection and refraction, we come to diffraction. Here the evidence was provocative to adherents of the corpuscular theory but still inconclusive for either theory of light. The physicist Francesco Grimaldi had discovered diffraction before Newton came on the scene, but his work was not published until 1665, two years after his death. In his experiments, Grimaldi allowed a beam of light into a darkened room through a small circular aperture and then passed it through a second aperture and onto a screen. He noticed that the spot of light on the screen was slightly larger than the second aperture, and that it had colored fringes. When he placed a thin obstacle in the beam, its shadow was not absolutely sharp: there were bright bands, very narrow and colored, following the outer edge of the shadow. In other words, light could be bent, or diffracted, by apertures and obstacles, contra Newton, even if the bending was only very slight. Newton repeated Grimaldi's experiments with the edge of a knife and also with a needle and observed colored fringes too on either side of the obstacle. It left him somewhat perplexed but still adhering to the corpuscular view. The third query in the last part of *Opticks* reads: "Are not the rays of light in passing by the edges and sides of bodies, bent several times backward and forwards, with a motion like that of an eel? And do not the three fringes of colored light above-mentioned arise from three such bendings?" Newton believed that the edges of an aperture or an obstacle interfered with the paths of the corpuscles, and he rejected the notion that diffraction was a wave phenomenon caused by a light wave spreading out from an aperture or bending around an obstacle. Although the wave offers the simpler of the two explanations of diffraction, it was nevertheless not obvious how the wave theory could account for the colors of the fringes.

We now have some idea of the scientific context of Young's first publication on light, "Sound and light", which was read to the Royal Society in January 1800 (before "On the mechanism of the eye"), shortly after his return to London from Cambridge, although it was actually written in Cambridge in mid-1799. Young had been led to the subject through his Göttingen dissertation on the human voice, through experiments on fluids such as blowing smoke through long tubes (which the Emmanuel College tutor had chanced upon), and even through his love of music. Introducing the paper, he wrote: "the further I have proceeded, the more widely the prospect of what lay before me has been extended; and … I find that the investigation, in all its magnitude, will occupy the leisure hours of some years, or perhaps of a life." The paper's section headings bear out his claim and give us a flavor of his many-sided approach to subjects of grand significance. They are:

I. The measurement of the quantity of air discharged through an aperture. II. The determination of the direction and velocity of a stream of air proceeding from an orifice. III. Ocular evidence of the nature of sound. IV. The velocity of sound. V. Sonorous cavities. VI. The degree of divergence of sound. VII. The decay of sound. VIII. The harmonic sounds of pipes. IX. The vibrations of different elastic fluids. X. The analogy between light and sound. XI. The coalescence of musical sounds. XII. The frequency of vibrations constituting a given note. XIII. The vibrations of chords. XIV. The vibrations of rods and plates. XV. The human voice. XVI. The temperament of musical intervals.

The most important section is section ten, on light and sound. Young begins by paying tribute to Newton's "incomparable writings" on optics, and then notes "one or two difficulties in the Newtonian system, which have been little observed." How, asks Young, can we account for the uniform velocity of light, if it is corpuscular?

> How happens it that, whether the projecting force is the slightest transmission of electricity, the friction of two pebbles, the lowest degree of visible ignition, the white heart of a wind furnace, or the intense heart of the sun itself, these wonderful corpuscles are always propelled with one uniform velocity? For, if they differed in velocity, that difference ought to produce a different refraction.

And how could the Newtonian system account for "a still more insuperable difficulty", the simultaneous refraction and reflection of light we have just discussed. "Why, of the same kind of rays, in every circumstance precisely similar, some should always be reflected, and others transmitted, appears in this system to be wholly inexplicable." Then there were Newton's rings to explain, that is, the colors produced by thin films. "The phenomena of [these] colors require, in the Newtonian system, a very complicated supposition, of an ether, anticipating by its motion the velocity of the corpuscles of light, and thus producing the fits of transmission and reflection; and even this supposition does not much assist the explanation." Finally, Young observes: "The greatest difficulty in this system is, to explain the different degree of refraction of differently colored light, and the separation of white light in refraction". The corpuscular system appeared to require a very uncertain number of differently colored corpuscles—an idea that seemed inherently improbable.

The most pregnant comment in the paper came in section eleven, however, on the coalescence of musical sounds. Young was offended by a suggestion of the prominent Cambridge mathematician Robert Smith in his *Harmonics* that two different sound waves could cross while remaining totally independent of each other. Smith's view implied that the two sounds at the point of crossing could somehow agitate the same air particles at the same time in two different ways. To Young this was a physical impossibility for two waves: "undoubtedly they cross, without disturbing each other's progress; but this can be no otherwise effected than by each particle's partaking of both motions." The two waves must *interfere* with each other, Young insisted. As proof, he cited the well-known phenomenon of *beats*. When two musical notes of similar loudness but slightly discrepant frequency are sounded at the same time, they produce a note of intermediate frequency that pulsates ('beats') in loudness. The more closely one note is tuned to the other, the slower the beat, until theoretically it disappears when the notes are of identical frequency. (A fact that is useful to piano tuners.) The cause of the beat, as Young establishes, must be the interference of the two sound waves, successively reinforcing and opposing each other as they move in and out of phase: the intermediate 'beating' note is loudest when the crests of the two sound waves coincide and quietest when the crest of one wave coincides with the trough of the other. The beat could not occur if it were possible for a crest and a trough to exist in the same air space at the same time, as maintained by Smith.

In his 1800 paper, Young does not take the next logical step and look for an optical equivalent of beats, but beats must surely have set him thinking about the possibility of the interference of light as well as sound— assuming light was a wave. Sir John Herschel, the physicist and astronomer (and photographic pioneer who coined the words negative, positive and snapshot), who became one of the early convinced advocates of the wave theory, imagined Young's mental process in 1800 after thinking about beats in sound, in a vivid letter to Hudson Gurney written 30 years later:

> What, then, is the analogous phenomenon in light? Can two lights destroy each other and produce darkness? If a class of phenomena in optics, referable to this or a similar principle, do exist, and could be made evident, [they] would afford a most cogent argument in support of the Huygenian doctrine ... Such, it is not unfair to conjecture, might be the train of thought which, arising in his mind while "sitting at his writing table", opened to him a new and vast field of experimental enquiry. Means of verification equally ready and simple were at hand. A scrap of card, a hair, and a candle; a few scratches on a bit of glass held in the sun and turned slowly around; a piece of paper, a pinhole, and a closed window shutter, [were] all the apparatus he required, and proved abundantly sufficient to satisfy himself and everyone else of the truth of a physical law so elegant as to command universal attention, and so important as at once to change the face of optical science.

There is poetic license in this—in reality, the wave theory took a long time to persuade most physicists—but the spirit of Herschel's remarks is true to Young's approach.

The revelation occurred in May 1801, according to Young himself, "while reflecting on the beautiful experiments of Newton". It so happens that this was exactly the time when he moved house from Norfolk Street to Welbeck Street. Perhaps—or is this too fanciful?—his domestic upheaval in some way aided his revolutionary optical reflections. (Curiously, Einstein's breakthrough in special relativity occurred around the time *he* moved house a century later.) By late June, as we know from Young's earlier letter to Dalzel, he was hard at work in Welbeck Street on the evidence for the undulatory theory. In July, he wrote a letter to *Nicholson's Journal* on another matter, which included the statement that "Light is probably the undulation of an elastic medium" and a point-by-point rationale for believing this. At the end, he noted that "all the phenomena of the colors of thin [films], which are in reality totally unintelligible on the common hypothesis"—that is, Newton's idea of fits—"admit a very complete and simple explanation by this supposition." A more detailed exposition of the undulatory theory was promised soon, "affording, from Newton's own elaborate experiments, a most convincing argument in favor of this system."

The promised paper, the writing of which must have gone hand in hand with Young's all-consuming preparation for his lecture course at the Royal Institution, was read to the Royal Society in November 1801 under the title, "On the theory of light and colors". We have come across it already in Chapter 5, "Physician of Vision", as the place of first announcement of Young's theory of three-color vision. Now we come to its announcement of the principle of interference of light.

Young starts by offering a hostage to fortune. He must have been only too keenly aware of Newton's famous caution concerning hypotheses, especially hypotheses about the nature of light; nevertheless he boldly states:

> Although the invention of plausible hypotheses, independent of any connection with experimental observations, can be of very little use in the promotion of natural knowledge; yet the discovery of simple

and uniform principles, by which a great number of apparently heterogeneous phenomena are reduced to coherent and universal laws, must ever be allowed to be of considerable importance towards the improvement of the human intellect.

Four hypotheses follow, the first three of them buttressed with chunks of quotation from Newton's writings, mainly from his *Opticks*. Here they are:

1. A luminiferous ether pervades the universe, rare and elastic in a high degree.

2. Undulations are excited in this ether whenever a body becomes luminous.

3. The sensation of different colors depends on the different frequency of vibrations excited by light in the retina.

4. All material bodies have an attraction for the ethereal medium, by means of which it is accumulated within their substance, and for a small distance around them, in a state of greater density, but not of greater elasticity.

Newton's authority could not be adduced in support of the fourth hypothesis because the hypothesis stood in direct opposition to Newton's own view. Young was suggesting that the ether was *denser within matter* than within space, while Newton thought the ether was rarer within matter. Young admits that he is unsure of this and that the fourth hypothesis is not "fundamental", unlike the other three; and within two years he abandoned his "ether distribution hypothesis" (the name given to it by Geoffrey Cantor) as unnecessary to explain the phenomena of light, without, however, abandoning the ether itself.

Note that there is no mention of the interference principle in the hypotheses. In Young's mind, it was already too well established an idea to label as a hypothesis. He includes it instead—without as yet using the word interference—under a list of nine "propositions" and their corollaries. Proposition eight reads: "When two undulations, from different origins, coincide either perfectly or very nearly in direction, their joint effect is a combination of the motions belonging to each." He explains why: "Since every particle of the medium is affected by each undulation, wherever the directions coincide, the undulations can proceed no otherwise than by uniting their motions, so that the joint motion may be the sum or difference of the separate motions, accordingly as similar or dissimilar parts of the undulations are coincident."

Having referred to the application of the principle to sound and beats, he then boldly writes: "it will appear to be of still more extensive utility in explaining the phenomena of colors". And he goes on (in the corollaries of proposition eight) to use interference to explain the colors of Newton's rings and the iridescent colors of "striated surfaces", such as some insect wings and integuments and mother-of-pearl.

But before coming to that explanation, let us have Young's clearest statement of all about his most far-reaching discovery. Instead of the phenomena of light or beats in sound, he chooses water waves to explain interference, because they can be simply visualized (the italics are mine):

> Suppose a number of equal waves of water to move upon the surface of a stagnant lake, with a certain constant velocity, and to enter a narrow channel leading out of the lake. Suppose then another similar cause to have excited another equal series of waves, which arrive at the same channel, with the same velocity, and at the same time with the first. Neither series of waves will destroy the other, but their effects will be combined: if they enter the channel in such a manner that the elevations of one series coincide with those of the other, they must together produce a series of greater joint elevations; but if the elevations of one series are so situated as to correspond to the depressions of the other, they must exactly fill up those depressions, and the surface of the water must remain smooth; at least I can discover no alternative, either from theory or from experiment.

> *Now I maintain that similar effects take place whenever two portions of light are thus mixed, and this I call the general law of the interference of light.*

Interference effects in water are easily demonstrated. Young included them in his Royal Institution lectures in 1802 using a device of his own invention, now known as a ripple tank, which was subsequently used by Faraday and soon became standard issue for lecturers in physics. The basic principle is to keep the water being agitated in a trough with a glass bottom and to illuminate the trough from below so that the water waves and their patterns cast shadows onto a white screen above the trough, as shown in Figure 7.3. In the published lectures, there is a famous diagram (see Figure 7.4) showing the interference of two sets of ripples. Young's caption reads as follows: "Two equal series of waves, diverging from the centers A and B, and crossing each other in such a manner, that in the lines tending towards C, D, E, and F, they counteract each other's effects, and the water remains nearly smooth"—what we now call *destructive* interference—"while in the intermediate spaces it is agitated"—*constructive* interference. You can see such interference patterns, with care, if you drop two stones of equal size into a pond at the same instant and watch their spreading ripples.

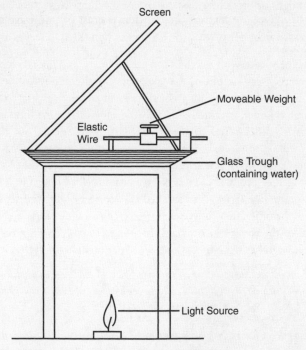

Figure 7.3   *Ripple tank for demonstrating wave interference, as visualized by Young in his* Natural Philosophy *(originally shown in a slightly different form).*

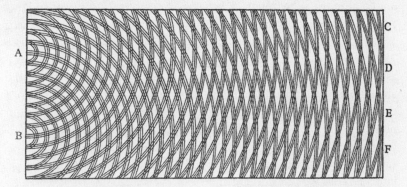

Figure 7.4 *Interference of two water waves, as shown in* Young's Natural Philosophy.

Constructive interference occurs when a crest of one wave coincides with a crest of another wave. This happens when the two waves are exactly in step, and also when they are out of step by exactly one, two or any integral number of wavelengths. Destructive interference occurs, conversely, when a crest coincides with a trough. This happens when the two waves are exactly out of step by half a wavelength, or by one-and-a-half, two-and-a-half or any half-integral number of wavelengths. Change the wavelength—by agitating the water faster or slower—and naturally the positions of constructive and destructive interference will alter.

With light, Young realized, constructive and destructive interference would produce patterns of alternating bright and dark, rather than areas of agitated and smooth water. However, the position of these patterns would be different for different colors, because, as he had hypothesized, color depended on wavelength. In reality, Newton's theory notwithstanding, there were no red and blue corpuscles; instead, there were light waves of longer and shorter wavelength. Here, Young saw, lay the correct explanation of the colors of Newton's rings.

Recall that in Newton's experiment with the lens and the glass plate, light rays are reflected from two surfaces, that of the lens-air interface at the top of the air film and of the air-glass plate interface at the bottom of the air film (see Figure 7.5). Using the wave theory and the principle of interference, Young saw that the colors in the rings must be the result of constructive interference between the light rays of each particular color reflected from the two surfaces. In the innermost red ring, red is reinforced by constructive interference and the other colors are diminished by destructive interference because the ray reflected from the bottom of the air film travels exactly one red wavelength further than the ray reflected from the top of the film. (With the second red ring, the ray in question travels two wavelengths further than the other ray, and so on for successive red rings.) Since all the distances between the bottom of the lens and the glass plate could be precisely calculated from the curvature of the lens, Young now had a way to measure the wavelength of red light. He gives it as 0.0000256 inches, which is very close to its modern accepted value. Newton's data enabled Young to calculate the wavelengths of the seven major colors in the visible spectrum.

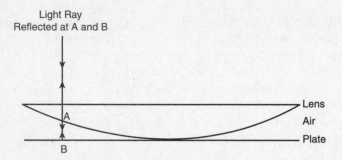

Figure 7.5   *Color in Newton's rings result from constructive interference between light reflected at point A and point B.*

Interference effects explain the colors of soap films too. Light is reflected from the front and the back of the film, and of course the highly visible colors, caused by constructive interference, depend on the exact thickness of the film. Polished surfaces with fine scratches—Young's "striated surfaces"—also display interference; rays of light are reflected from adjacent scratches and constructively interfere, producing the colors. We have already mentioned the iridescence of natural substances. A modern synthetic example is the laser-read compact disc, which displays a rainbow of colors that vary as one turns it in ordinary daylight, depending on the precise angles of reflection of light from the grooves in the disc's surface.

Young's 1801 paper launched his undulatory theory and the principle of interference of light. But it did not provide the kind of incontrovertible evidence of interference required to persuade fellow scientists. He did not yet have the optical equivalent of water waves crossing in a ripple tank or beats between musical notes. His explanations of colors, though profoundly ingenious, were not the only possible interpretations (not least, there was Newton's own). He needed to demonstrate interference unambiguously; to find an experiment that was open to only *one* convincing theoretical interpretation—the wave interpretation.

Two years later, after finishing his lectures at the Royal Institution, Young finally came up with what he had been looking for. In November 1803, he announced it in his lecture, "Experiments and calculations relative to physical optics". "In making some experiments on the fringes of colors accompanying shadows, I have found so simple and so demonstrative a proof of the general law of the interference of two portions of light, which I have already endeavored to establish, that I think it right to lay before the Royal Society a short statement of the facts that appear to me so decisive." Without more ado, he describes the experiment (I have italicized the crucial part):

> I made a small hole in a window shutter, and covered it with a piece of thick paper, which I perforated with a fine needle. For greater convenience of observation I placed a small looking-glass without the window shutter, in such a position as to reflect the sun's light, in a direction nearly horizontal, upon the opposite wall, and to cause the cone of diverging light to pass over a table on which were several little screens of card paper. I brought into the sunbeam a slip of card, about one-thirtieth of an inch in breadth, and observed its shadow, either on the wall or on other cards held at different distances. Beside the fringes of color on each side of the shadow, the shadow itself was divided by similar parallel fringes, of smaller dimensions, differing in number, according to the distance at which the shadow was observed, but leaving the middle of the shadow always white. *Now these fringes were the joint effects of the portions of light passing on each side of the slip of card, and inflected, or rather diffracted, into the shadow. For, a little screen being placed a few inches from the card, so as to receive either edge of the shadow on its margin, all the fringes which had before been observed in the shadow on the wall, immediately disappeared,* although the light inflected on the other side was allowed to retain

its course, and although this light must have undergone any modification that the proximity of the other edge of the slip of card might have been capable of occasioning. ... Nor was it for want of a sufficient intensity of light that one of the two portions was incapable of producing the fringes alone; for, when they were both uninterrupted, the lines appeared, even if the intensity was reduced to one-tenth or one-twentieth.

From his measurements of diffraction of light with cards, Young calculated the wavelengths of different colors and compared his values with the wavelengths he had earlier calculated from Newton's rings, and he also compared his own measurements of the diffraction fringes with those of Newton using a knife edge and a needle. The results were compelling evidence that these diffraction phenomena and the colors of Newton's rings were due to interference, and that light was indeed a wave. "The foundations of the wave theory of light had been well and truly laid," wrote Alex Wood, Young's biographer, looking back from the 1950s. In 1803, however, almost no one immediately accepted Young's radical conclusions. Newton's aura was powerful indeed, and one of his acolytes was deeply disturbed by Young's undulations. We shall now see how this critic almost succeeded in strangling the infant theory at birth.

Chapter 8

'Natural Philosophy and the Mechanical Arts'

"[The] phenomena of nature resemble the scattered leaves of the Sibylline prophecies; a word only, or a single syllable, is written on each leaf, which, when separately considered, conveys no instruction to the mind; but when, by the labor of patient investigation, every fragment is replaced in its appropriate connection, the whole begins at once to speak a perspicuous and a harmonious language."

Young, introduction to A Course of Lectures on Natural Philosophy and the Mechanical Arts, *1807*

On 14 June 1804, the day after his 31st birthday, Young got married. His bride, Eliza Maxwell, was youthful in age, even by the standards of the time, only 18 or 19 years old; but the marriage proved to be a happy one, although there were no children. We do not know how the two of them first met, but since the Maxwells kept a house in Cavendish Square, a mere quarter of a mile from Welbeck Street, Thomas and Eliza were practically neighbors. She was the second daughter of James Primrose Maxwell, Esq., of Trippendence, near Farnborough in the county of Kent, who was himself a member of the younger branch of the family of Sir William Maxwell of Calderwood Castle in Lanarkshire, a county in south-west central Scotland. It appears that Young was still smitten with the memories of the aristocratic 'goddesses' who had charmed him in the wilds of the Scottish Highlands on his horseback tour nearly ten years earlier, while he was a student in Edinburgh.

Young's biographer George Peacock, who knew Eliza Young for a long time and eventually yielded to the "affectionate constancy" with which she urged him to write about her late husband, has very little to say about the relationship. "It was a marriage of mutual affection and esteem, such as he had always looked forward to as the great object of his professional and other exertions, and secured him a home which was graced by all the refinements of good manners and a cultivated taste: it was a singularly happy marriage." This brevity might be regarded simply as conventional Victorian piety and reticence about matrimony, except that there are a few hints from others (which we shall come to in their proper place) that Eliza was concerned with Thomas's work; also, it is very clear from Young's letters, quoted by Peacock, that he was really fond of his wife's three sisters, especially Emily, for whom he penned his autobiographical sketch. Writing in that, Young says of his marriage that it "was happy, though without the comforts of a family: comforts which, for a great part of his life would have been accompanied by deep anxiety, and of which the absence was in great measure supplied by other domestic affections." From the context of his remarks, the "deep anxiety" was almost certainly a reference to financial worries: the remarks follow some comments about his income that suggest that his lack of success as a medical practitioner was a source of both professional and marital concern. A few years later, having at last secured an important hospital position after a hard-fought election, in one of his rare surviving references to his wife, Young told his friend Hudson Gurney: "Mrs Young has emerged from death to life by the event of this contest."

Young's twentieth-century biographers, Frank Oldham and Alex Wood, without Peacock's advantage of personal acquaintance with Mrs Young, her sisters, Gurney and others in their subject's immediate circle—and with most of Young's papers having disappeared—were not in a position to add anything of substance to Peacock's picture of the marriage. Nevertheless, Oldham made the reasonable conjecture from the limited evidence available that Eliza Young was of definite help in keeping her husband productive and unembittered as a scientist and scholar: "Misjudged, incompletely understood, unfairly treated by many of those scientists who had chosen to further the knowledge of those subjects which he had so highly endowed by his genius, it needed a devoted and patient understanding wife to help him weather the storms of criticism and abuse which would have soured most men."

Oldham was almost certainly thinking of the libelous attack on Young's theory of the interference of light that appeared in the very recently launched but already highly influential *Edinburgh Review* about three months after his marriage, in October 1804. But it is also true that for the rest of his life, Young would suffer from lack of appreciation, misunderstanding and some open hostility in all his major fields of activity— physics, physic and Egyptology—as well as in his scientific work for the Government. Periodically he hit back, but for the most part he complained only in private or kept silent.

The first attack in the *Edinburgh Review* appeared in its second number in January 1803, and consisted of a review of Young's paper, "On the theory of light and colors", read to the Royal Society in 1801 and published in its volume of *Philosophical Transactions* for 1802, and of a second review of a second paper by Young on the same subject, also published in the same volume. As with the criticism published in 1804, which related to Young's 1803 Royal Society lecture, "Physical optics", these reviews were all unsigned.

The tone of the attack, one of zestful sarcasm, was set right at the beginning. The first review opens:

> As this paper contains nothing which deserves the name, either of experiment or discovery, and, as it is in fact destitute of every species of merit, we should have allowed it to pass among the multitude of those articles which must always find admittance into the collections of a society which is pledged to publish two or three volumes every year. The dignities of the author, and the title of Bakerian Lecture, which is prefixed to these lucubrations, should not have saved them from a place in the ignoble crowd. But we have of late observed in the physical world a most unaccountable predilection for vague hypothesis daily gaining ground ... We wish to raise our feeble voice against innovations, that can have no other effect than to check the progress of science, and renew all those wild phantoms of the imagination which Bacon and Newton put to flight from her temple. ... Has the Royal Society degraded its publications into bulletins of new and fashionable theories for the ladies who attend the Royal Institution? *Proh pudor*! Let the Professor continue to amuse his audience with an

endless variety of such harmless trifles; but, in the name of science, let them not find admittance into that venerable repository which contains the works of Newton, and Boyle, and Cavendish, and Maskelyne, and Herschel.

By the time of the third review, some two years later, the sarcasm has given way to abuse:

In our second number, we exposed the absurdity of this writer's "law of interference", as it pleases him to call one of the most incomprehensible suppositions that we remember to have met with in the history of human hypotheses. ... The long silence [of the author] led us to flatter ourselves, either that he had discontinued his fruitless chase after hypotheses, or that the [Royal] Society had remitted his effusions to the more appropriate audience of both sexes which throngs round the chairs of the Royal Institution. The volume now before us, however, at once destroys all such expectations. The paper which stands first, is another Bakerian Lecture, containing more fancies, more blunders, more unfounded hypotheses, more gratuitous fictions, all upon the same field on which Newton trod, all from the fertile, yet fruitless, brain of the same eternal Dr Young.

Young had ignored the first and second reviews, at least in print, but this third one was too malignant to overlook. Not only was his intellect being ridiculed, his character too was being impugned, by open suggestions that he changed his scientific views whimsically and performed experiments incompetently—all of which Young understandably thought might damage his standing with the general public as a junior physician whose medical reputation was by no means established. In late November 1804, he therefore wrote a lengthy response, entitled "Reply to the animadversions of the Edinburgh reviewers", and published it as a pamphlet, as was the custom of the time, priced one shilling.

In a few passages, he gives as good as he gets from the reviewer. For example, on the intellectual argument, Young writes:

Conscious of [his] inability to explain the [diffraction] experiment which I have advanced, too ungenerous to confess that inability, and too idle to repeat the experiment, he is compelled to advance the supposition that it was incorrect, and to insinuate that my hand may easily have erred through a space so narrow as one-thirtieth of an inch. But the truth is, that my hand was not concerned: the screen was placed on a table, and moved mechanically forwards with the utmost caution ... and I assert that it was as easy to me to estimate an interval of one-thirtieth of an inch, as an interval a hundred or a thousand times as great. Let him make the experiment, and then deny the result if he can.

And on the tone of the attack, he writes:

[T]he writer confesses that he has not "sufficient fancy to discover" how the "interference of two portions of light" could ever produce an appearance of color. The poverty of his fancy may indeed easily be admitted, but it is unfortunate that he either has not patience enough to read, or intellect enough to understand, the very papers that he is criticizing; for, if he had perused with common attention my Bakerian Lecture on light, he might have understood such a production of color without any exertion of fancy at all.

But on the whole, Young restrains himself and restricts his reply to questions of natural philosophy. There was a deep scientific issue at stake—the same as had divided Newton and Huygens—and at least a few worthwhile intellectual criticisms, mixed with the invectives of the reviewer. Notably, the latter's distrust of hypotheses in science and his vaunting of experimental evidence versus Young's predilection for hypotheses and his relative lack of enthusiasm for experiment. This debate continues even today, in that non-scientists generally imagine that scientists first do experiments and then look for hypotheses to explain the results, whereas in fact, normally, hypothesis determines experiment, which then acts as a check on the hypothesis. Young explains this cogently in his reply:

[T]here are two general methods of communicating knowledge; the analytical, where we proceed from the examination of effects to the investigation of causes; the other synthetical, where we first lay down the causes, and deduce from them the particular effects. In the synthetical manner of explaining a new theory we necessarily begin by assuming principles, which ought, in such a case, to bear the modest name of hypotheses; and when we have compared their consequences

with all the phenomena, and have shown that the agreement is perfect, we may justly change the temporary term *hypothesis* into *theory*.
This mode of reasoning is sufficient to attach a value and importance
to our theory, but it is not fully decisive with respect to its exclusive
truth, since it has not been proved that no other hypothesis will agree
with the facts. It is exactly in this manner that I have endeavored to
proceed in my researches.

Young also 'outs' the reviewer in his reply, naming him as Henry
Brougham, one of the founding contributors of the *Edinburgh Review*.
Brougham was then in his mid-twenties and yet to make his name as a
politician; in due course he would become Lord Brougham, lord chancel-
lor of England, a noted educationist and a celebrated maverick of the
Georgian and Victorian ages. (He gave his name to the brougham, the
horse-drawn closed carriage with the driver perched outside in front.)
Since Brougham never denied Young's charge, and continued to believe in
the corpuscular theory of light as late as the 1850s—when the Royal
Society finally refused to publish any more of his outdated views—there
can be no question about his authorship of the three reviews attacking
Young in 1803-04.

Brougham's motives are somewhat murkier. He surely disagreed profoundly with Young's undulatory theory, but he had a personal grudge too.
In 1800, Young had written slightingly (if accurately) of one of Brougham's
mathematical papers, and Brougham wanted his revenge. But even before
this, in 1795, in a letter written in his mid-teens to Sir Charles Blagden, one
of the secretaries of the Royal Society, Brougham had criticized Young's
first paper for the Royal Society, "Observations on vision", "which I cannot
help flattering myself without vanity is neither better grounded nor more
new than my own." (A criticism he stingingly reworked in his first review
of Young's papers.) Blagden, we may recall, was the gossip who had started
off the rumor of plagiarism against Young two years earlier and then had
to withdraw it. Significantly, Blagden encouraged Brougham, and helped
to have his work published by the Royal Society in 1796 and again in 1797,
when Brougham was not yet 20. But Brougham was not elected a fellow
until well after Young, despite his strong desire to join this scientific elite;
and then it was more for his social connections than for his scientific
work, which never amounted to other than dilettantism. It seems only too

probable that Brougham perceived the polymathic Young to be a scientific rival, whom he envied—a Mozart to his Salieri.

How much real damage his reviews did to Young's theory is perhaps more questionable. A wounded Young signed off his pamphlet—which appears not to have sold a single copy—by stating emphatically: "With this work my pursuit of general science will terminate: henceforwards I have resolved to confine my studies and my pen to medical subjects only." Peacock (who published his biography while Brougham was still alive) was in no doubt of the damage, writing that "it would be difficult to refer to another example where the irresponsible power of anonymous criticism has been so unscrupulously exercised, or where the effects which it produced were so long and so injuriously felt." Peacock claimed that the attacks delayed the serious examination of Young's theories in Britain "for nearly twenty years". Lord Rayleigh, writing in 1889, well after Brougham's death, commented that, "It is doubtless true that Young's views did not at the time of publication of [his Royal Institution] lectures command the authority which now attaches to them", and footnoted this comment laconically, without any further explanation· "I may remark, in passing, that Brougham knew a little of experimenting, as of everything else, except law!"—which plainly implies that Rayleigh assumed that all his scientific readers would be aware of the significance of Brougham's attack on Young. However, some modern scholars think that Brougham's deleterious impact was overrated in the nineteenth century and point instead to undoubted flaws in Young's presentation of his theory of interference as the main reason for the long delay in its acceptance by the physicists of his day. While this is probably true, we cannot doubt that Brougham's coruscating criticism must have undermined Young's credibility among the reading public, not to mention his self-confidence.

There is a letter about the impact written by one of Young's friends, George Ellis, a minor writer and historian, to his close friend Sir Walter Scott. Ellis had spread Young's pamphlet around and was anxious to see justice done to him. According to Ellis, a bookseller who had agreed to pay Young £1000 for the right to publish his Royal Institution lectures came to Welbeck Street and told him that the "ridicule" of his work in the *Edinburgh Review* "had so frightened the whole [book] *trade* that he must request to be released from his bargain." It might even have been this incident, which would have been enough to rattle any author, that prompted Young to write his pamphlet with its anguished final statement about giving up "general science".

Needless to say, he did not give up his non-medical studies, though he did begin his long retreat into publishing anonymously. Physic would continue to take second place in his life to physics (so to speak), until the publication of his lectures in 1807. But from 1804 onwards, Young started to build up a medical practice on the south coast not far from Brighton in the small seaside resort of Worthing, a place then fashionable with those who could not visit the continent because of the war with France. Over the next fifteen years or so, for about four months every summer, during the season for sea bathing, Young and his wife would leave London and live in sunny Worthing. In 1808, they bought a house there and in due course Dr Young became known as Worthing's "resident physician". (It was in Worthing, as we shall see, that he first made progress with the Rosetta Stone and the Egyptian hieroglyphs.)

Young's *A Course of Lectures on Natural Philosophy and the Mechanical Arts* is a magnificent-looking work, consisting of two quarto volumes running to more than fifteen hundred pages, with a plate section containing color illustrations in addition to some fine black-and-white engravings. Even the exacting Young was pleased, despite the fact that he never received a penny from the publisher because the firm went out of business just as the book appeared. The first volume consists of the lectures and their illustrations; the second includes Young's papers not delivered as Royal Institution lectures (such as "On the mechanism of the eye"), and a unique catalog of the scientific literature from the ancient Greeks up to about 1805 with extensive commentary by Young. Of this catalog, which is organized in relation to the content of the lectures, Young writes: "the labor of arranging about twenty thousand articles, in a systematic form, was by no means less considerable than that of collecting them. The transactions of scientific societies, and the best and latest periodical publications, which have so much multiplied the number of the sources of information"—how twenty-first century that observation sounds!— "constituted no small part of the collection, which was thus to be reduced into one body of science." Only Young, perhaps, among the scientists of his day (or, *a fortiori*, our day), would have had the command of foreign languages, combined with the range, judgment and industry to compile such a monumental bibliography.

Gurney—who admittedly never claimed scientific expertise—writing in 1830, just after Young's death, called Natural Philosophy "a mine to which every one has since resorted, [which] contained the original hints of more things since claimed as discoveries, than can perhaps be found in a single production of any known author." He recalled that "one of the men most distinguished for science in Europe has been known to say, that if his library were on fire, and he could save only one book from the conflagration, it should be the lectures of Dr Young."

A generation later, the professor of mathematics at Edinburgh University, Philip Kelland, the editor of the second edition of Natural Philosophy, wrote in his preface:

> Whether we regard the depth of Dr Young's learning, the extent of his research, the accuracy of his statements, or the beauty and originality of his theoretical views, in whatever way we contemplate these lectures, our admiration is equally excited. ... Unlike other popular writers, who, for the most part, either take the sciences at second hand, or content themselves simply with extracting the discoveries and adopting the hypotheses of more distinguished philosophers, Dr Young traveled over the whole literature of science, and whilst we are astonished at the rich store of materials which he has collected, we find nothing more prominent than the impress of his own acute and powerful mind.

A century later, the physicist Sir Joseph Larmor, who was Lucasian professor of mathematics at Cambridge (the position once held by Newton and now held by Stephen Hawking), described Natural Philosophy as "the greatest and most original of all general lecture courses" in a substantial article on Young published in Nature. Of the second volume, Larmor wrote: "No such authoritative catalog, even of the select classical works of modern science, of personal origin, is likely to appear again."

Finally, in our own time, when the lectures were republished in 2002, Nicholas Wade, professor of visual psychology at Dundee University, introduced the volumes with the comment: "Reprinting them renders Young's insights accessible to modern scientists, who will marvel that one mind could encompass so much. ... The weakness of Young's verbal delivery contrasted starkly with the strength of the written one, and it is the latter that has stood the test of time and warrants wonder."

These four tributes make obvious why it is well nigh impossible to do full justice to the lectures in *Natural Philosophy* in a mere few pages. We must simply pick out a handful of ideas from them that were influential or prescient. At the head of the list must come Young's most renowned experiment of all, the one that begins this book, in which he used two narrow slits to split a beam of light into two beams and observe their interference fringes on a screen. This was not part of his November 1803 lecture to the Royal Society (attacked by Brougham), which as we know described an experiment showing a single beam of light diffracted by a narrow card. Young must therefore have conducted the double-slit experiment during the period after 1803 and before 1807—perhaps partly in response to the criticism of Brougham. He was naturally searching for an experiment with light that would be as persuasive and definitive as the experiments showing the interference of two water waves or two sound waves (in the form of beats).

In *Natural Philosophy*, he describes the key experiment with the double slits as follows:

> [T]he simplest case appears to be, when a beam of homogeneous light falls on a screen in which there are two very small holes or slits, which may be considered as centers of divergence, from whence the light is diffracted in every direction. In this case, when the two newly formed beams are received on a surface placed so as to intercept them, their light is divided by dark stripes into portions nearly equal, but becoming wider as the surface is more remote from the apertures, so as to subtend very nearly equal angles from the apertures at all distances, and wider also in the same proportion as the apertures are closer to each other. The middle of the two portions is always light, and the bright stripes on each side are at such distances, that the light, coming to them from one of the apertures, must have passed through a longer space than that which comes from the other, by an interval which is equal to the breadth of one, two, three, or more of the supposed undulations [i.e., one, two, three or more wavelengths], while the intervening dark spaces correspond to a difference of half a supposed undulation, of one and a half, of two and half, or more.

And he illustrates this with the drawing shown in Figure 8.1, which he captions: "The manner in which two portions of colored light, admitted through two small apertures, produce light and dark stripes or fringes by their interference, proceeding in the form of hyperbolas; the middle ones are however usually a little dilated, as at A".

Figure 8.1   *Double-slit experiment and interference fringes, as shown in Young's Natural Philosophy—his most celebrated discovery.*

It appears definitive—and there is no question that the double-slit experiment does demonstrate the interference of light, as countless others have subsequently shown. But did Young actually perform it? Or was it only a 'thought' experiment, like Einstein's notion of trying to catch up with a light ray? At least one current historian of science, John Worrall, thinks the latter was the case: Young's double-slit experiment was an intuition of the truth, not a real experiment. Worrall bases his view on the following undoubted facts: Young does not explicitly state that he did the experiment; Young provides no numerical data; Young says nothing about the light source he used and the other experimental conditions; and Young never again refers to the experiment.

A second historian, Nahum Kipnis, disagrees with Worrall's conclusion, yet Kipnis thinks that Young did not interpret his experimental observations correctly. After carefully examining all of Young's statements on the subject in the published lectures, he concludes:

> Young did experiment with two slits, and he used both white and monochromatic light. However he did not discover the interference fringes: he confused them with diffraction fringes. Because the interval between the slits was too large, he could see only diffraction fringes produced by each slit separately. One of the obvious reasons for his mistake was the qualitative character of his observation: if he had measured the distance between the observed fringes, he would have immediately realized that they were of the wrong kind.

The debate is a subtle one. No one doubts that it was feasible in Young's day to have observed interference fringes with a candle, two suitably made slits and a screen. Young was a practical man, more than competent to construct apparatus. It seems inherently improbable that he did not perform this relatively straightforward experiment, as alleged by Worrall, given its great significance for his theory; and most unlikely that he did not report faithfully what he saw, given his integrity in all his other work. But it is certainly surprising that he gives no numerical data about the fringes, unlike, say, his calculations of the wavelengths of different colors based on Newton's rings. And if his results were as conclusive as he described, one would have expected him to report them more prominently than he did by burying them in the middle of a huge mass of other work. But then perhaps he could not face further skepticism and even hostile attacks on his theory. We should also recall that he now wanted to stay out of the limelight and concentrate on becoming a physician.

Whatever the truth may have been, this episode illustrates the fine balance in science between theory and experiment. Young knew what he expected to see on the basis of his wave theory, and he saw it demonstrated in his experiment. As so often happened with him, his theory was correct, even if in this case his experimental results may not have shown what he imagined them to. Better-designed experiments subsequently proved his theory right, and that is what ultimately matters. But having said this, we can also see how his lack of experimental precision was part of the reason why Young's theory would take a long time to triumph over the reigning corpuscular theory.

A second key part of the published lectures deals with mechanics. Young grasped the importance of what physicists would later term the kinetic energy of a moving body, and he was the first person (as pointed out by James Clerk Maxwell) to use the term *energy* in its modern scientific sense—as a measure of a system's ability to 'do work'. When we throw a stone vertically upwards, physicists say that the stone exchanges its kinetic energy—i.e., its energy of motion—for potential energy as it does *work* against the force of gravity; at the top of its trajectory, when the stone is momentarily at rest, all its kinetic energy has been transformed into potential energy. Because of gravity, the higher a stone is from the ground, the more potential energy it has. The same is basically true, too, if we throw the stone at an angle, so that its trajectory is curved; again it exchanges the kinetic energy of the vertical component of its motion for potential energy. In his lecture, "On confined motion", Young writes:

> [S]ince the height, to which a body will rise perpendicularly, is as the square of its velocity, it will preserve a tendency to rise to a height which is as the square of its velocity, whatever may be the path into which it is directed, provided that it meets with no abrupt angle … The same idea is somewhat more concisely expressed by the term energy, which indicates the tendency of a body to ascend or to penetrate to a certain distance, in opposition to a retarding force.

In the lecture, "On collision", Young goes further and defines this energy as the mass of a body multiplied by the square of its velocity:

> The term energy may be applied, with great propriety, to the product of the mass or weight of a body, into the square of the number expressing its velocity. Thus, if a weight of one ounce moves with a velocity of a foot in a second, we may call its energy 1, if a second body of two ounces [has] a velocity of three feet in a second, its energy [2] will be twice the square of three, or 18.

Thus, Young expresses the idea that if the same body moves twice as fast, its energy is not twice as great but four times as great, and if it moves three times as fast, its energy is nine times greater. Hence the fact—crucial to safe drivers—that the stopping distance of a car is proportional not to its speed but to the *square* of its speed. Today, in classical physics (in other words, when physicists are able to ignore Einstein's small relativistic correction), a body's kinetic energy is defined in the same way as Young did in 1807, with one comparatively trivial refinement.

Another mechanics lecture deals with elasticity—what Young calls "passive strength". In his preface, he writes: "The passive strength of materials of all kinds has been very fully investigated, and many new conclusions have been formed respecting it, which are of immediate importance to the architect and to the engineer". Later in life, he would apply this thinking to practical problems such as the building of ships and bridges.

It is here that Young defines what all engineers now know as Young's modulus of elasticity. This specifies how different materials contract or stretch when compressed or extended. Engineers need the modulus to calculate the compression and extension of beams under stress. For the physicist, the modulus is defined as the ratio of the applied stress to the resulting strain, where the stress is the compressive force per unit area of cross-section and the strain is the compression per unit length. Young's own definition was as follows:

> [W]e may express the elasticity of any substance by the weight of a certain column of the same substance, which may be denominated the modulus of its elasticity, and of which the weight is such, that any addition to it would increase it in the same proportion, as the weight added would shorten, by its pressure, a portion of the substance of equal diameter.

As Alex Wood frankly notes: "This definition is a model of cumbersome obscurity." Even so, Young's concept was right.

He also studied how materials bend and shear. A current textbook on the mathematical theory of elasticity states: "Young was the first to introduce shear as a form of elastic strain, and observed that the resistance of a body to shear is different than to extension and compression." However, Young did not introduce a separate shear modulus, as is now used by engineers and physicists; his modulus applies only to longitudinal stress and strain.

Stress and strain and shear naturally led him to think about the forces that hold materials together, manifested in their tensile strength, that is their resistance to breaking under tension. In 1807, the atomic nature of matter was much more of a hypothesis than a theory; in fact, atoms and molecules were not totally accepted as real entities by most physicists until the turn of the nineteenth century and the work of Einstein on Brownian motion. Young, as usual, was ahead of the pack. (Another was the chemist John Dalton, with his concept of atomic weights of chemical elements, introduced in 1808.) In fact, Young was the first physicist to make an experimental estimate of the diameter of a molecule, based on his study of capillary action and surface tension in liquids—about fifty years before the estimates of Lord Kelvin. Surface tension accounts for the meniscus in water in a glass capillary tube, for the considerable force required to separate two plates of glass sandwiching a drop of water, and for an insect's inability to fly when its wings are thoroughly wet. It was Lord Rayleigh who first pointed out Young's achievement from his study of Young's lectures around 1890, as Rayleigh's son explains in his biography of his father: "Tearing a liquid column in half .., creates two surfaces. These surfaces have a tension, and Young showed that the range of molecular forces could be found by comparing the surface tension with the tensile strength." Young estimated the extent of the cohesive molecular forces as being limited to about 250 millionths of an inch and assumed that this should be taken as the limit of a molecule. He later wrote of the "ultimate atoms of bodies, of water, for instance, about a million of which would occupy a length equal to the diameter of one of the red particles of blood." His molecular estimate was about a hundred times too large, but his method and reasoning were acute.

When he turned his attention to heat, Young wholeheartedly embraced the avant-garde idea that heat was the manifestation of atomic motion, and rejected the dominant idea that it was a substance—an imponderable fluid then called *caloric*—which was said to increase within a body the hotter it was. Young was aware of and appreciated Count Rumford's recent experiments on heat, which had begun to undermine the caloric theory—this was one of the reasons why Rumford had appointed Young to lecture at the Royal Institution—but again he was well ahead of his time. Then he went further and linked heat and light as one phenomenon in a blindingly clear-sighted passage in *Natural Philosophy*. Here Young proposed the modern concept of a continuous spectrum of radiation, passing from invisible ultraviolet through visible light to invisible infrared, with the wavelength increasing and the frequency decreasing:

> If heat is not a substance, it must be a quality; and this quality can only be motion. It was Newton's opinion, that heat consists in a minute vibratory motion of the particles of bodies, and that this motion is communicated through an apparent vacuum, by the undulations of an elastic medium, which is also concerned in the phenomena of light. If the arguments which have lately been advanced, in favor of the undulatory nature of light, be deemed valid, there will be still stronger reasons for admitting this doctrine respecting heat, and it will only be necessary to suppose the vibrations and undulations, principally constituting it, to be larger and stronger than those of light, while at the same time the smaller vibrations of light, and even the blackening rays [ultraviolet light], derived from still more minute vibrations, may, perhaps, when sufficiently condensed, concur in producing the effects of heat. These effects, beginning from the blackening rays, which are invisible, are a little more perceptible in the violet, which still possess but a faint power of illumination; the yellow green afford the most light; the red give less light, but much more heat, while the still larger and less frequent vibrations [infrared light], which have no effect on the sense of sight, may be supposed to give rise to the least refrangible rays, and to constitute invisible heat.

Here we see Young doing what all great scientists try to do: unify as many phenomena of the physical world as possible within an all-embracing theoretical structure. Of course he failed in many respects, especially with electricity and magnetism. But it is amazing that in 1807 he managed to achieve as much as he did, guided by his broad and deep knowledge of the theories and experiments of both earlier and contemporary scientists, by his own experiments and, most important of all, by his formidable intuition. As a slightly exasperated Peacock remarks of Young's writings:

> Important and difficult steps are passed over as manifest, terms are neglected as insignificant, analogies take the place of proofs, and we are surprised to find ourselves at the end of an investigation, even

within the limits of space which would commonly be deemed hardly
sufficient to master the difficulties which meet us at the beginning.
But his rare sagacity hardly ever deserts him.

The preface of *Natural Philosophy* is dated 30 March 1807. Almost six
years had gone by since Young was appointed as professor of natural philosophy at the Royal Institution and began to prepare his lectures. They
had been years of enormous intellectual challenge and excitement, of
backbreaking toil, of acerbic public criticism and searing private anxiety,
along with domestic contentment since his marriage. Yet Young knew that
the two volumes contained work of enduring importance, even if they
were unlikely to be fully appreciated in his lifetime. Along with his papers
for the Royal Society, included in the second volume, the lectures in
Natural Philosophy would most probably be his epitaph. But a book on
natural philosophy, however great, could do little to further his medical
career and a reliable income. The time had now come, Young knew, to
focus on medicine, or forever forgo his claims as a physician.

Chapter 9

Dr Thomas Young, M.D., F.R.C.P.

"There is no study more difficult than that of physic: it exceeds, as a science, the comprehension of the human mind."

Young, Introduction to Medical Literature, *1813*

In transferring his main attentions from natural philosophy to physic, Young moved from a science into a world that remained much closer to an art than a science. So very little was known in 1807 about so much of the workings of the human body (even the stethoscope had yet to be invented), that physicians were often hardly more medically competent than their patients. As Young candidly confessed to his intimate friend Gurney in 1806, "in consultations, however opposite opinions may be, it is usual to tell the patient that the parties consulted are perfectly agreed, and it is very unfair to examine the witnesses separately where so much depends on opinion."

He gives a prime example of medical practice from his personal experience:

> I was dining at the duke of Richmond's one day last winter, and there came in two notes, one from Sir W. Farquhar, and the other from Dr Hunter, in answer to an inquiry whether or no his grace might venture to eat fruit pies or strawberries. *I trembled for the honor of the profession*, and could not conceal my apprehensions from the company: luckily, however, they agreed tolerably well, the only difference of opinion being on the subject of pie-crust.

His idea now was to apply his mind to the medical literature, going right back to the classical writers Hippocrates and Galen, written in as many languages as he could muster, and to try to sort out the wheat from the chaff, rather as he had just finished doing in his 20,000-article catalog for the second volume of *Natural Philosophy*. He told Gurney in another letter, written early in 1807:

> I purpose seriously to do something in physic, by collecting all that is worth knowing, and comparing it with the general economy of the operations of nature. I do not know who has attempted to do this soberly: [Erasmus] Darwin had neither patience nor precision enough; and I am confident that much more may be learnt and taught in this way than from a routine of old woman's practice, which is all that a fashionable physician obtains. In many other departments of science I have been enabled to draw conclusions from a comparison of the experiments of others, which I should have been much longer in discovering by investigations of my own; *and why not in physic?*

Why not indeed? In 1809-10, Young gave a course of lectures at London's Middlesex Hospital. This led to his publishing, in 1813, *An Introduction to Medical Literature, Including a System of Practical Nosology* (nosology is the study of the classification of diseases), which went into a second edition in 1823. It included a description of a valuable new invention, an instrument that Young dubbed an eriometer, useful for "the measurement of minute particles, especially those of blood and of pus". Two years later, in 1815, he followed with *A Practical and Historical Treatise on Consumptive Diseases*, a book on a subject of personal relevance given his own narrow escape from consumption as a teenager in the late 1780s. These two books were remarkable surveys, incorporating both his own growing knowledge as a practitioner and his cullings from two millennia of medical writings. But if Young imagined that they would help to bring him honor among physicians and success with patients, as it seems he did, he was doomed to disappointment. Professional reviews mixed considerable praise with some severe criticism. Overall, it would not be exaggerating much to say that Young's medical books disturbed his colleagues by demonstrating the profession's ignorance and failures, and probably discouraged patients by adding to Dr Young's reputation for being a cold man of science. "Certainly, science was not what secured top clinicians their fame," comments the historian Roy Porter on this period of medicine in *The Greatest Benefit to Mankind*. "The public regarded him as over-wise!" Young's ever-concise French physicist friend Dominique Arago wrote after his death.

The early nineteenth century was a time when quackery was endemic throughout medicine, even at the highest levels of the profession, where certain celebrated physicians were willing to lend their names to ineffective patent medicines. It was mainly to expose such quackery, and also the nepotism rife in hospital appointments, that today's leading British medical journal, *The Lancet*, was founded in 1823 by Thomas Wakley. Porter summarizes the situation frankly in his *Quacks: Fakers and Charlatans in Medicine*:

> Medical men of all sorts were competing for custom, recognition, and reward. Each in his own way—top physician, humble general practitioner, empiric, folk healer—made his bid to seize the moral high ground in a medical arena in which the law was acknowledged to be dog-eat-dog. It is revealing that most of the anecdotes passed down to us about pre-modern practitioners center, admiringly, on their love of lucre and their success in getting it. ... With such cash-conscious competition (the ambience against which Victorian professional medicine reacted), it is the similarities rather than the differences between quacks and regulars that deserve to be highlighted.

A much-quoted piece of eighteenth-century doggerel about John Coakley Lettsom, the famous Quaker physician from the generation before Young's, who died in 1815, ran as follows:

When any sick to me apply

I physics, bleeds and sweats 'em,

If, after that, they choose to die,

Why verily,

I. Lettsom.

The unpalatable truth was that the only eighteenth-century improvements in practical medicine with a proven record of saving lives were in the prevention of smallpox: first by inoculation—practiced by Young's own physician Thomas Dimsdale in the 1760s—and then by vaccination—the treatment pioneered by John Hunter's former pupil Edward Jenner in the late 1790s. Every other kind of medical treatment was to a great extent hit-or-miss.

Young recognizes this truth with typical rationality and honesty in his book on consumptive diseases:

> [I]t is probable that without assistance not one case in 1000 of the disease would recover; and with the utmost power of art, perhaps not more than one in 100 will be found curable. However discouraging this representation may be on the one hand, it is still some consolation, supposing it correct, to think, on the other hand, that ten times as many lives may be saved by medical treatment as without it: and we may be induced, by this statement, to argue with extreme caution respecting the comparative value of the medicines which we may think proper to prescribe: for since the utmost, that we can expect from the operation of the most powerful remedies, is to save one out of 100 cases of confirmed consumption, we must have witnessed the failure of any new mode of treatment in at least 50 cases, before we are fully authorized to suppose, that it has been less successful than the most effectual remedies previously known ...

Hence, of course, the proliferation of remedies, each claiming at least one inspiring success. Without plumping for any one regimen—such was the uncertainty surrounding this dread disease—Young's *Consumptive Diseases* examines in detail

> the respective advantages of bleeding, whether general or local; of cathartics, as neutral salts, calomel, and sulfur; of emetics, as ipecacuan, tartar emetic, sulfate of zinc, and sulfate of copper; of sorbefacients, as digitalis, mercury, and alkalis; of epispastics, as blisters, issues, caustics, cauteries, and setons; of sudorifics, as antimony, Dover's powder, and sarsaparilla; of expectorants, as gum ammoniac, squills, and polygala; of demulcents, as oils and gums; of narcotics, as opium and hemlock; of suppuratories, or detergents, as balsams and

balsamic vapors; of astringents, as the mineral and vegetable acids, cathechu and kino; of tonics, as steel, myrrh, bark, Angustura, and lichen; of diet of various kinds; of exercise; and of a change of climate.

Of emetics, which were intended to make the patient vomit, Young notes: "In my own case, small doses of tartar emetic appeared to be of use in relieving the hectic symptoms, for a few weeks, at an early period, taken in an oily mixture, with which it made a nauseating compound." With patients who had reached the stage of haemoptysis, the expectoration of blood or bloody mucus—which Young fortunately had not as a teenager— he recommends a combination of ipecacuan, the root of a South American shrub, and sulfate of soda, taken every four hours.

With such a combination, I have relieved several cases of haemopty-sis and of internal haemorrhage more speedily than by any other means, not excepting even the acetate of lead, which seems also to be a less permanent, as well as a less safe remedy. In the case of Mrs K., the wife of a tradesman in Welbeck Street, the haemoptysis was accompanied by hectic fever, and an expectoration apparently puru-lent: she had lately been confined, and there was no symptom which encouraged me, at first sight, to think favorably of the probable result; but the mixture appeared almost immediately to diminish the quantity of blood expectorated, and it agreed so well, that she contin-ued to take it every four hours, for some months, without alteration; the expectoration and the febrile symptoms gradually subsided, and she has remained in perfect health for more than a year.

However Young adds ominously: "The same combination of medi-cines was, for a time at least, equally successful in a case of haemoptysis, which afterwards occurred at Worthing; but the final event yet remains to be determined." Did this consumptive patient recover too? We shall never know.

It is tempting to quote further cases, not least because Young writes well about medicine, probably better, generally speaking, than he does about natural philosophy; medicine's unavoidable imprecision and its involvement with the humanities and the classical languages seem to have liberated his pen. But I shall resist. We cannot leave his book, though, without quoting from the historical survey of consumption, which forms the second part of the book (as in *Natural Philosophy*, current thinking comes first, not history). Here are three characteristically interesting cases he reports, ranging from the serious to the bizarre.

Young cites a Latin work, published in Venice in 1761, by the celebrated eighteenth-century anatomist, Giovanni Battista Morgagni, the pioneer of the post-mortem examination. A patient in Lucca with a laryngeal phthisis, but with little or no fever, was treated by confining him to a warm but spacious room, forbidding him from speaking to his friends except in a whisper, and feeding him "about half a pint of milk from the breast every morning and afternoon, and puddings made of barley meal, ginseng and milk," while "enjoining abstinence from wine." Having followed this regimen from November to May, the patiently completely recovered. Now, every person in Lucca suffering from consumption adopted the same diet—"but without a single additional instance of success."

Then there is the treatise in French of a forgotten physician named Raulin, published in Paris in 1784, for which Young has rather less respect. It is "full of accounts of great cures, and of implicit faith in the doctrines of Hippocrates: of [Raulin's] reasoning powers we may judge, by his theory of the mechanism of respiration, which he attributes to the expansion of the portion of air inhaled, by the effect of heat, in consequence of which it is again expelled by the denser air rushing in". Young comments sardonically: "it seems that many people succeed in physic, who could not have made a pair of bellows."

Finally, Young derides a work by the pathologist Bonetus (Théophile Bonet), also in Latin, published in Geneva in 1706:

> The compilation of Bonetus is considered as one of the classical works upon the morbid changes produced by diseases: but it is encumbered with numberless repetitions, and a very unnecessary multiplication of authorities: some of the histories are ludicrously marvelous, but none more so than those which are copied from an anonymous work on watercresses. A surgeon of Brussels cursorily advises a consumptive countryman to live on watercresses, raw and boiled: after a year, to his great surprise, the man returns perfectly cured: the surgeon takes him into a private room, under the pretence

of examining him more minutely, and stabs him with a stiletto, in order to satisfy his curiosity, with respect to the state of his lungs, which are again regenerated: the man's wife, who has been waiting for her husband, suspects some mischief, and gives information to a magistrate: but the surgeon finally obtains a pardon, on account of his great skill.

Young became an M.D. in 1808 and a fellow of the Royal College of Physicians, F.R.C.P, in December 1809. But what he needed most to establish himself was an appointment as a physician to a London hospital. Although such positions were not in themselves financially remunerative, they provided invaluable experience of patients, conferred prestige on the physician, and led to opportunities for private practice. Young entered himself as a candidate at the Middlesex Hospital in 1806 but withdrew after discovering that there was a favored internal candidate. By lecturing at the hospital three years later, he hoped to ingratiate himself there. But not until January 1811 was he able to secure a position, and then it was not at the Middlesex but at St George's Hospital.

Located on Hyde Park Corner, close to his late great-uncle's house off Park Lane, St George's had been founded some eighty years earlier, in 1733, by a group of philanthropists, specifically to treat the 'deserving' poor. It was known as John Hunter's hospital—he became a surgeon there in 1768, having formerly been a pupil too—although his relationship with his colleagues was always tempestuous. In the 1790s, the hospital passed through a fallow patch, but it recovered, and by 1811, a position there as a physician was considered extremely desirable for someone aspiring to a major private practice.

Young, with no connections at St George's, was up against two other strong candidates. Influential friends such as Gurney lobbied for him, and he was also active on his own behalf. The outcome of the election, on 24 January, was 100 votes for Young, 92 for Dr Cabbell, 51 for Dr Roget (who became the author, in the 1850s, of the famous *Thesaurus*) and 4 for Dr Harrison. Young wrote to Gurney just afterwards:

[A]ny of the three candidates had advantages which would have secured him in any common case. Local interest and the protracted efforts of a whole family made the Cabbells very naturally confident of triumphant success; parliamentary influence and the natural wish to serve a man who is likely to be lord chancellor, made Sir S. Romilly's nephew [Dr Roget] very formidable; and for myself the event speaks. But it is remarkable what a variety of interests I have been obliged to bring into play; scarcely any one of my friends having procured for me more than two or three favorable answers, so that every one lamented how very little he could do; yet the aggregate was sufficient for the purpose.

His wife Eliza, as mentioned earlier, obtained a new lease of life from her husband's appointment.

Young did not become a pillar of St George's, but neither did he fight tooth and nail with his colleagues as Hunter had. He says nothing about the place in his autobiographical sketch, other than mentioning the bare fact of his appointment "after a very arduous contest". And there is no record at all of his name in the minutes of the hospital governors' meetings until he died eighteen years later in 1829, when a curt notice appears: "the death of Dr Young, one of the physicians to this institution, having been announced, *ordered* that a special court for the election of a physician to this institution in the place of Dr Young, deceased, be held ..."

On the other hand, Young was always attentive to his duties at the hospital. Even after he retired from private practice in the very late 1810s, he remained a physician at the hospital and continued to do his regular round of the patients there. The apothecary of the hospital is said to have remarked that more of Young's patients were cured, or at least relieved of pain, than the patients of other physicians who used more vigorous and fashionable treatments. There are references to cases at St George's in Young's medical books, and he makes sure to mention his position there on the books' title pages. After his death, his portrait was hung in the board room of the hospital.

According to Young's biographer Peacock, it was a current observation among the pupils at the hospital—shades of the Cambridge students who called him "Phenomenon" Young—that "Dr Young was a great philosopher, but a bad physician." Peacock comments: "the credence which was given to it, both within and without the walls of the hospital, continued to give strength to the very prejudice in which it originated—that the highest professional eminence [in medicine] is only attainable by the exclusion of all other pursuits."

Young was acutely cognizant of this attitude among physicians, while at the same time despising it as unworthy of the subject. One of his long letters to his sister-in-law Emily, written in 1815, includes a nonsense verse, on a Latin model, which touches on his divided attitude to his chosen profession:

Medical men, my mood mistaking,

Most mawkish, monstrous messes making,

Molest me much; more manfully,

My mind might meet my malady:

Medicine's mere mockery murders me.

(He then jokes to Emily: "This is bad enough; but it is much more difficult to execute such alliterations in English than in Latin, and you will say it would be no great loss if it were impossible.") In that last line, one senses a deliberate ambiguity: that medical men are mocking him, while medicine is mocking them.

Peacock was strongly in sympathy with Young's dislike of some aspects of his profession, perhaps because he himself had suffered from bad medical advice during a period of serious illness that almost derailed his biography. No doubt thinking of some mannered and opinionated top physicians he had known, Peacock writes:

It is the peculiar misfortune of the medical profession that its members can rarely dare to confess their ignorance, thinking it more or less necessary—in order to maintain their influence with their patients and with the world—to speak with equal decision, whether they are authorized by their knowledge to do so or not ... The real fact is that the prestige of a reputation once attained, whether through the influence of charlatanism, good fortune, or superior merit, is not easily destroyed, and the very eccentricities and extravagances which repel patients of sense and delicacy, tend to confirm the prepossessions of those who are wanting in these qualities, and who are naturally apt to wonder at or admire what they do not understand.

These remarks, not too surprisingly, incensed a very senior medical man, and at the same time provoked him to some interesting comments on Young. Sir Benjamin Brodie, ten years Young's junior, was a surgeon at St George's in his time as a physician there, who in fact carried out the post-mortem examination of Young's body in 1829. In due course, Brodie became sergeant-surgeon to King William IV and Queen Victoria, president of the Royal College of Surgeons, the first president of the General Medical Council, and finally president of the Royal Society—the first surgeon to hold this office. In his autobiography, published posthumously (and also after the death of Peacock) in 1865, Brodie writes:

> Nothing can be more unjust than the whole of Dr Peacock's observations on this subject. There may be among physicians, as well as in other professions, some individuals who acquire a reputation to which they have no claim, but my experience justifies me in asserting that no physician acquires a *large* reputation, or retains what may be called an extensive practice, who is really unworthy of it. The public are, on the whole, pretty good judges in a matter in which they are so much interested, and if by any accident they have been led to give their confidence to a wrong person, they are seldom long in discovering and correcting their mistake.

Then Brodie comes to Young:

> The truth is that either his mind, from it having been so long trained by the study of the more exact sciences, was not fitted for the profession which he had chosen, or that it was so much engrossed with other, and to him more interesting pursuits, that he never bestowed on it that constant and patient attention without which no one can be a great physician or a great statesman. The students at the hospital complained that they learned nothing from him. I never could discern that he kept any written notes of cases, and I doubt whether he ever thought of his cases in the hospital after he had left the wards. His medical writings were little more than compilations from books, with no indication of original research. I offer these observations as a matter of justice to others, and not in depreciation of Dr Young, for whom I had a great personal regard, whose vast and varied attainments out of his profession, and whose great original genius displayed in other ways, place him in the foremost rank of those whose names adorn the annals of our country.

Here speaks a conventional medical mind (quoted by the standard *Roll of the Royal College of Physicians of London* in its entry for Young, written by William Munk). Amidst the huffing and puffing of a top surgeon, the usual institutional accusations are being leveled against Young, more or less like those of the Emmanuel College tutor at Cambridge, if we omit the by then fairly standard encomium for Young's non-medical work. Young does not focus enough to succeed; he is not involved with his fellow professionals; he does not seem to do any solid research. The first and second accusations had an element of truth in them, but the accusation about lack of originality is nonsense, and particularly rich coming from Brodie, given that he was one of those present, as a young surgeon at St George's, when Young and Everard Home witnessed the testing of the seafarer Benjamin Clerk, one of whose eyes had lost the power to accommodate after a cataract operation (as discussed in Chapter 5, "Physician of Vision"). The outcome was a triumph for Young's brilliant experimental work on accommodation, and a defeat for Home. We can reasonably assume that the notoriously unscrupulous Home, who was Brodie's sponsor at St George's, would not have had the friendliest of feelings toward Young, which Home would no doubt have had plenty of opportunity to communicate to his pupil Brodie. Anyway, some two centuries on, the fact is that Young's physiological research is remembered, while the medical research of Brodie—notwithstanding his roll-call of professional honors—is forgotten by doctors, and even by the medical reference books.

This is not to imply that Young was as significant in the history of medicine as he was in the more exact sciences. Most of his avowedly medical writing (in other words, excluding his work on the eye and vision) was in the nature of a discriminating survey of the works of others, though he did give a decidedly original lecture to the Royal Society in 1808, "On the functions of the heart and arteries". Apart perhaps from his invention of the eriometer—which was taken up and developed by a later generation of physiologists—Young certainly has no claim to be a leading physician of his age, in the class of Edward Jenner or Louis Pasteur. Perhaps the fairest assessment of his medical work comes from the surgeon Joseph Pettigrew's 1839 portrait of Young in his *Biographical Memoirs of the Most Celebrated Physicians, Surgeons, etc. Who Have Contributed to the Advancement of Medical Science*. Pettigrew writes:

[Young] was not a popular physician. He wanted that confidence or assurance which is so necessary to the successful exercise of his profession. He was perhaps too deeply informed, and therefore too sensible of the difficulty of arriving at true knowledge in the profession of medicine, hastily to form a judgment; and his great love of and adherence to truth made him often hesitate where others felt no difficulty whatever in the expression of their opinion. He is therefore not celebrated as a medical practitioner; nor did he ever enjoy an extensive practice; but in information upon the subjects of his profession, in depth of research into the history of diseases, and the opinions of all who have preceded him, it would be difficult to find his equal.

For Young himself, there can be no doubt, especially from his many disappointed letters to Gurney, that his relative lack of success as a physician became a sore point. He avoids the issue of his medical reputation almost entirely in his autobiographical sketch. He continued to hope for major professional success until around 1815, with the appearance of his two medical books. Thereafter, his unparalleled range of other interests took over and gradually pushed medicine into the background. He was now about to enter yet another new world of knowledge: the strange universe of ancient Egypt.

Chapter 10

Reading the Rosetta Stone

"You tell me that I shall astonish the world if I make out the inscription. I think it on the contrary astonishing that it should not have been made out already, and that I should find the task so difficult as it appears to be."

Young, letter to Hudson Gurney, 1814

Young became hooked on the scripts and languages of ancient Egypt in 1814, the year he began to decipher the Rosetta Stone. He continued to study them with variable intensity for the rest of his life, literally until his dying day. The challenge of being the first modern to read the writing of what appeared then to be the oldest civilization in the world—far older than the classical civilization of his beloved Greeks—was irresistible to a man who was as equally gifted in languages, ancient and contemporary, as he was in science. He himself in 1823 described his obsession as being driven by "an attempt to unveil the mystery, in which Egyptian literature has been involved for nearly twenty centuries". His epitaph in London's Westminster Abbey states, accurately enough, that Young was the man who "first penetrated the obscurity which had veiled for ages the hieroglyphics of Egypt"—even if it was Jean-François Champollion who in the end would enjoy the glory of being the first actually to read the hieroglyphs.

But before we delve into Young's attempt, we need some historical background, a survey of earlier thinking of the kind in which Young himself specialized (for example, in his *Natural Philosophy* and his *Consumptive Diseases*). To decipher the Egyptian hieroglyphs in the period

1814-24 required Young and Champollion to sweep away centuries of erroneous thinking, dating back to classical antiquity, while building on a handful of genuine insights from various scholars.

The civilization of the pharaohs had gone into eclipse some two thousand years earlier, when it was conquered by Alexander the Great and came under the Hellenistic rule of the Ptolemaic dynasty. Such was its legendary magnificence, however, that the Greeks and Romans, especially the Greeks, regarded ancient Egypt with a paradoxical mixture of reverence for its wisdom and antiquity and contempt for its 'barbarism'. The very word hieroglyph derives from the Greek for 'sacred carving'. Egyptian obelisks were taken to ancient Rome and became symbols of prestige; today, thirteen large obelisks stand in Rome, while only four remain in Egypt.

The classical authors generally credited Egypt with the invention of writing (though Pliny the Elder attributed it to the inventors of cuneiform). But none of them could read the hieroglyphs in the way that they were able to read the Greek and Latin alphabet, despite the fact that hieroglyphic inscriptions continued to be written in Egypt as late as AD 394. They preferred to believe, as Diodorus Siculus wrote in the first century BC, that the Egyptian writing was "not built up from syllables to express the underlying meaning, but from the appearance of the things drawn and by their metaphorical meaning learned by heart." In other words, the hieroglyphs were conceptual or symbolic, not phonetic like their own alphabet. Thus, a hieroglyphic picture of a hawk represented anything that happened swiftly; a crocodile symbolized all that was evil.

By far the most important authority was an Egyptian magus named Horapollo (Horus Apollo) supposedly from Nilopolis in Upper Egypt. His treatise, *Hieroglyphika*, was probably composed in Greek, during the fourth century AD or later, and then sank from view until a manuscript was discovered on a Greek island in about 1419 and became known in Renaissance Italy. Published in 1505, the book was hugely influential: it went through 30 editions, one of them illustrated by Albrecht Dürer, and even remains in print.

Horapollo's readings of the hieroglyphs were a combination of the (mainly) fictitious and the genuine. Young called them "puerile ... more like a collection of conceits and enigmas than an explanation of a real system of serious literature". For instance, according to the esteemed *Hieroglyphika*:

> [W]hen they wish to indicate a sacred scribe, or a prophet, or an embalmer, or the spleen, or odor, or laughter, or sneezing, or rule, or judge, they draw a dog. A scribe, since he who wishes to become an accomplished scribe must study many things and must bark continually and be fierce and show favors to none, just like dogs. And a prophet, because the dog looks intently beyond all other beasts upon the images of the gods, like a prophet. ...

and so on. There are elements of truth in this: the jackal ("dog") hieroglyph writes the name of the god Anubis, who is the god of embalming, a smelly business (hence the meaning "odor"?); and a recumbent jackal writes the title of a special type of priest, the 'master of secrets', who would have been a sacred scribe and considered something of a prophet; while a striding jackal can also stand for an official, and hence perhaps for a judge. But consider Horapollo's "What they mean by a vulture":

> When they mean a mother, or boundaries, or foreknowledge ... they draw a vulture. A mother, since there is no male in this species of animal. ... The vulture stands for sight since of all other animals the vulture has the keenest vision. ... It means boundaries, because when a war is about to break out, it limits the place in which the battle will occur, hovering over it for seven days. Foreknowledge, because of what has been said above [about sight] and because it looks forward to the amount of corpses which the slaughter will provide it for food.

This was almost all fantasy—except for "mother": the hieroglyph for mother is indeed a vulture.

The Arabs who occupied Egypt with the coming of Islam in the medieval period had a marginally more accurate understanding of the hieroglyphs because they at least believed that the signs were partly phonetic, not purely symbolic. (Their attribution of phonetic values was wrong, however.) But this belief did not pass from the Islamic world to the European. Instead, fuelled by Horapollo, the Renaissance revival of classical learning brought a revival of the Greek and Roman belief in the hieroglyphs as symbols of wisdom. The first of many scholars in the modern world to write a whole book on the subject was a Venetian, Pierius Valerianus. He published it in 1556, and illustrated it with delightfully fantastic 'Renaissance' hieroglyphs.

The most famous of these interpreters was the Jesuit priest Athanasius Kircher. In the mid-seventeenth century, Kircher became Rome's accepted pundit on ancient Egypt. But his voluminous writings took him far beyond 'Egyptology'; "sometimes called the last Renaissance man" (notes the *Encyclopaedia Britannica*), and dubbed "the last man who knew everything" in a recent academic study, Kircher attempted to encompass the totality of human knowledge. The result was a mixture of folly and brilliance—with the former easily predominant—from which his reputation never recovered.

In 1666, Kircher was entrusted with the publication of a hieroglyphic inscription on an Egyptian obelisk in Rome's Piazza della Minerva. This had been erected on the orders of Pope Alexander VII to a design by the sculptor Bernini (it stands to this day, mounted on a stone elephant, encapsulating the concept 'wisdom supported by strength'). Kircher gave his reading of a *cartouche*—i.e., a small group of hieroglyphs in the inscription enclosed by an oval ring—as follows: "The protection of Osiris against the violence of Typho must be elicited according to the proper rites and ceremonies by sacrifices and by appeal to the tutelary Genii of the triple world in order to ensure the enjoyment of the prosperity customarily given by the Nile against the violence of the enemy Typho." Today's accepted reading of this cartouche is simply the name of a pharaoh, Wahibre (Apries), of the twenty-sixth dynasty!

An ironic Young—who was also, let us not forget, a Quaker by upbringing, with a natural resistance to priestly authority and tradition—noted of Father Kircher: "according to his interpretation, which succeeded equally well, whether he happened to begin at the beginning, or at the end, of each of the lines, they all contain some mysterious doctrines of religion or of metaphysics." Mumbo-jumbo had a ready market in seventeenth-century Rome, just as it had in ancient Rome and, indeed, in our twenty-first-century world.

By contrast, Kircher genuinely assisted in the rescue of Coptic, the language of the last phase of ancient Egypt, by publishing the first Coptic grammar and vocabulary. The word Copt is derived from the Arabic *qubti*, which itself derives from Greek *Aiguptos* (Egypt). The Coptic script was invented around the end of the first century AD, and from the fourth to the tenth centuries Coptic flourished as a spoken language and as the official language of the Christian church in Egypt; after that it was replaced by Arabic, except in the church, and by the time of Kircher, the mid-seventeenth century, the language was headed for extinction (though it was still used in the liturgy). During the eighteenth century, however, several scholars acquired a knowledge of Coptic and its alphabet, which in its standard form consists of the 24 Greek letters plus six signs borrowed from the last stage of the script of ancient Egypt (the demotic script, which appears on the Rosetta Stone along with the hieroglyphic script, as we shall shortly see). This knowledge of Coptic would prove essential in the decipherment of the hieroglyphs in the nineteenth century.

Wrong-headed theories about the Egyptian script flourished throughout the Enlightenment period. For example, a Swedish diplomat, Count Palin, suggested in three publications that parts of the Old Testament were a Hebrew translation of an Egyptian text—which was a reasonable conjecture—but then Palin tried to reconstruct the Egyptian text by translating the Hebrew into Chinese. This was not quite as crazy as it sounds, given that both Egyptian hieroglyphs and Chinese characters have a strong conceptual and symbolic element; the very fact of the existence of the Chinese script, and also a particular structural link between the two scripts, would in fact offer an important clue in deciphering Egyptian in the more cautious hands of others. But Palin went way too far with his hieroglyphic extravaganza. As Young noted coolly:

> [T]he peculiar nature of the Chinese characters ... has contributed
> very materially to assist us in tracing the gradual progress of the
> Egyptian symbols through their various forms; although the resem-
> blance is certainly far less complete than has been supposed by
> Mr Palin, who tells us, that we have only to translate the Psalms of
> David into Chinese, and to write them in the ancient character of that
> language, in order to reproduce the Egyptian papyri, that are found
> with the mummies.

The first 'scientific' step in the right direction came from an English clergyman. In 1740, William Warburton, the future bishop of Gloucester, suggested that all writing, hieroglyphs included, might have evolved from pictures, rather than by divine origin. Abbé J. J. Barthélemy, an admirer of Warburton, then made a sensible guess in 1762 that obelisk cartouches might contain the names of kings or gods—ironically, on the basis of two false observations (one being that the hieroglyphs enclosed in the oval rings differed from all other hieroglyphs). Finally, near the end of the eighteenth century, a Danish scholar, Georg Zoëga, hazarded that some hieroglyphs might be, in some measure at least, what he called "notae phoneticae", Latin for "phonetic signs", representing sounds rather than concepts in the Egyptian language. The path toward decipherment was at last being cleared.

And now we have reached a turning point: the arrival of Napoleon's invasion force in Egypt in 1798 and the discovery of the Rosetta Stone. The word cartouche, as applied to Egyptian hieroglyphs, dates from this fateful expedition. The oval rings enclosing groups of hieroglyphs, visible within inscriptions on temple walls and elsewhere in Egypt to any casual observer, reminded the French soldiers of the cartridges (*cartouches* in French) in their guns.

Fortunately, the military force was almost as interested in culture as in conquest. A party of French savants, including the celebrated mathematician Jean-Baptiste Fourier, accompanied the army and remained in Egypt for some three years. There were also many artists, chief of whom was Domenique Denon. Between 1809 and 1828, Denon and others illustrated the *Description de l'Égypte*, and the whole of Europe (especially a polymath like Young) was astonished by the marvels of the pharaohs. One of the French drawings shows the city of Thebes, with the columns of the temple of Luxor behind and highly inscribed obelisks in the foreground. The carved scenes depict the charge of chariot-borne archers under the command of Ramses II against the Hittites in the battle of Kadesh (*c.* 1275 BC). Napoleon's army was so awestruck by this unheralded spectacle that, according to a witness, "it halted of itself and, by one spontaneous impulse, grounded its arms."

It was a demolition squad of French soldiers that stumbled across the Rosetta Stone in mid-July 1799, probably built into a very old wall in the village of Rashid (Rosetta), on a branch of the Nile just a few miles from the sea. Recognizing its importance, the officer in charge had the stone moved immediately to Cairo. Copies were made and distributed to the scholars of Europe during 1800—a remarkably open-minded gesture considering the politics of the period. In 1801, the stone was shifted to Alexandria in an attempt to avoid its capture by British forces. But after a somewhat unseemly wrangle, it was eventually handed over, shipped to Britain, and displayed in the British Museum, where it has remained ever since, apart from an excursion to Paris in October 1972 for the 150th anniversary of Champollion's decipherment.

According to one of the museum's curators of Egyptian antiquities, Richard Parkinson, the Rosetta Stone "is the most popular single artifact in the British Museum's collections". In his catalog of the exhibition "Cracking Codes", celebrating the bicentenary in 1999 of the stone's discovery, he writes: "Unfortunately, the stone's iconic status seems to encourage visitors to reach out and touch the almost miraculous object." The familiar white characters on the black surface, polished by generations of visitors' hands until the stone looked more like a printer's lithographic stone (which it was actually used as, in the early nineteenth century) than a two-thousand-year-old monument, were mainly the result of chalk and carnauba wax rubbed into the surface by museum curators to increase visibility and aid preservation. In the 1990s, in time for the bicentenary, this policy was changed and the stone cleaned to reveal its natural color. It is now seen to be a dark gray slab of igneous rock (not basalt, as formerly believed), which sparkles with feldspar and mica and has a pink vein through its top left-hand corner; it weighs some three quarters of a ton.

Even a quick glance reveals that the stone is broken—this fracture probably occurred before it came to Rosetta—both in the right-hand corner and, most obviously, at the top. So the inscription is incomplete. Fortunately, there exist other similar complete inscriptions (found after the decipherment), including a near-copy inscribed fourteen years later and now in the Cairo Museum, so we can visualize the Rosetta Stone as it would originally have looked (see Figure 10.1).

Figure 10.1 *Rosetta Stone, as it would have looked before it was broken (drawn by C. Thorne and R. Parkinson and reproduced with their permission).*

From the moment of discovery, it was clear that the inscription on the stone was written in three different scripts, the bottom one being the Greek alphabet and the top one—the most badly damaged—Egyptian hiero-glyphs with visible cartouches, as shown in Figure 10.2. Sandwiched between them was a script about which little was known. It plainly did not resemble the Greek script, but it seemed to bear at least a passing resem-blance to the hieroglyphic script above it, though without having any car-touches. Today we know this script as *demotic*, a development (*c.* 650 BC) from a cursive form of writing known as *hieratic* used in parallel with the hieroglyphic script from as early as 3000 BC (hieratic itself does not appear on the Rosetta Stone). The name derives from Greek *demotikos*, meaning 'in common use'—in contrast to the sacred hieroglyphic, which was essen-tially a monumental script. The term demotic was first used by Champollion, who refused to import Young's earlier coinage, *enchorial*, which Young had adopted from the description of the second script given in the Greek inscription: *enchoria grammata*, or 'letters of the country'.

The first step toward decipherment was obviously to translate the Greek inscription. This was done by, among others, Young's classicist friend Richard Porson and, more accurately, by C. G. Heyne, the professor Young had known at Göttingen in the 1790s. It turned out to be a decree issued at Memphis, the principal city of ancient Egypt, by a general coun-cil of priests from every part of the kingdom assembled on the first anniversary of the coronation of the young Ptolemy V Epiphanes, king of all Egypt, on 27 March 196 BC. Greek was used because it was the language of court and government of the descendants of Ptolemy, Alexander's gen-eral. The names Ptolemy, Alexander and Alexandria, among others, occurred in the Greek inscription.

Much of the decree is taken up, to put it bluntly, with the terms of a deal by which the priests agreed to give their support to the new king (who was only thirteen) in exchange for certain privileges. While this was of some interest to historians of ancient Egypt and its religion, the eye of would-be decipherers was caught by the very last sentence. It read: "This decree shall be inscribed on a stela of hard stone in sacred [i.e., hiero-glyphic] and native [i.e., demotic] and Greek characters and set up in each of the first, second and third [-rank] temples beside the image of the ever-living king." In other words, the three inscriptions—hieroglyphic, demotic and Greek—were equivalent in meaning, though not necessarily 'word for word' translations of each other. This was how scholars first knew that the stone was a bilingual inscription: the kind most sought after by decipherers, a sort of Holy Grail of decipherment. The two languages were clearly Greek and (presumably) ancient Egyptian, the language of the priests, the latter being written in two different scripts—unless the "sacred" and "native" characters concealed *two* different languages, which seemed unlikely from the context. (In fact, as we now know, the Egyptian languages written in hieroglyphic and demotic are not identical, but they are closely related, like Latin and Renaissance Italian.)

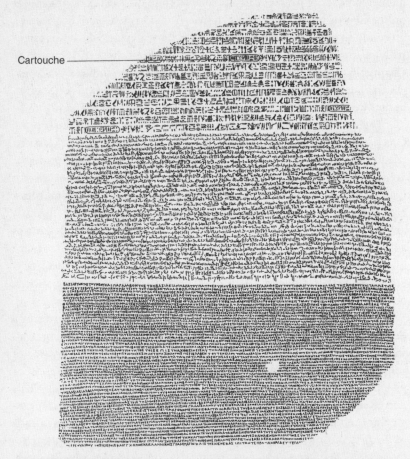

Cartouche

Figure 10.2   *Rosetta Stone inscription in hieroglyphic script (at top, with cartouche indicated), in demotic script (middle) and in Greek script (bottom).*

Since the hieroglyphic section was so damaged, attention focused on the demotic. Two scholars, a distinguished French Orientalist in Paris, Sylvestre de Sacy (a future teacher of Champollion), and a Swedish diplomat, Johan Åkerblad, adopted similar techniques. They searched for a name, such as Ptolemy, by isolating repeated groups of demotic symbols located in roughly the same position as the eleven known occurrences of Ptolemy in the Greek inscription. Having found these groups, they noticed that the names in demotic seemed to be written alphabetically, as in the Greek inscription—that is, the demotic spelling of a name apparently contained more or less the same number of signs as the number of letters in its assumed Greek equivalent. By matching demotic sign with Greek letter, they were able to draw up a tentative alphabet of demotic signs. Certain other demotic words, such as 'Greek', 'Egypt', and 'temple', could now be identified using this demotic alphabet. It looked as though the entire demotic script might be alphabetic like the Greek inscription.

But in fact it was not, unluckily for de Sacy and Åkerblad. Young was sympathetic: "[They] proceeded upon the erroneous, or, at least imperfect, evidence of the Greek authors, who have pretended to explain the different modes of writing among the ancient Egyptians, and who have asserted very distinctly that they employed, on many occasions, an alphabetical system, composed of 25 letters only." Taking their cue from classical authority, neither de Sacy nor Åkerblad could get rid of his preconception that the demotic inscription was written in an alphabetic script—as against the hieroglyphic inscription, which both scholars took to be wholly *non*-phonetic, its symbols expressing ideas, not sounds, along the lines of Horapollo. The apparent disparity in appearance between the hieroglyphic and demotic signs, and the suffocating weight of European tradition that Egyptian hieroglyphs were a conceptual or symbolic script, convinced them that the invisible principles of hieroglyphic and demotic were wholly different: the hieroglyphic had to be a conceptual/symbolic script, the demotic a phonetic/alphabetic script.

Except for one element. De Sacy deserves credit as the first to make an important suggestion: that the foreign names inside the hieroglyphic cartouches, which he naturally assumed were Ptolemy, Alexander and so on, were also spelt *alphabetically*, as in the demotic inscription. He was led to this by some information given to him by one of his pupils, a student of Chinese, in 1811. The Chinese script was at this time generally thought in Europe to be a primarily conceptual script like the hieroglyphs. Yet, as this student pointed out, foreign (i.e., non-Chinese) names, such as those of the Jesuit missionaries in China, had to be written *phonetically* in Chinese with a special sign to indicate that the Chinese characters were being reduced to a phonetic value without any conceptual value. (English-speakers, of course, indicate some foreign words in writing English with their own 'special sign'—italicization.) Were not Ptolemy, Alexander and so on, Greek names foreign to the Egyptian language, and might not the cartouche be the ancient Egyptian hieroglyphic equivalent of the special sign in Chinese? But as for the rest of the hieroglyphs—all those not enclosed in cartouches—de Sacy was convinced they must undoubtedly be non-phonetic.

Enter Young in earnest, in early 1814. One might have expected him to have involved himself earlier with the Rosetta Stone, after it first went on display in London in 1802. However, at that time, he was totally occupied with the Royal Institution lectures, and after the mammoth task of publishing these in 1807, he devoted himself mainly to medicine. What finally triggered his active interest in the hieroglyphs, he tells us, was a review he wrote in 1813 of a massive work in German on the history of languages, *Mithridates, oder Allgemeine Sprachkunde* by Johann Christoph Adelung, which contained a note by the editor "in which he asserted that the unknown language of the stone of Rosetta, and of the bandages often found with the mummies, was capable of being analyzed into an alphabet consisting of little more than thirty letters". When an English friend shortly afterwards returned from the East and showed Young some fragments of papyrus he had collected in Egypt, "my Egyptian researches began". First he examined the papyri and reported on them to the Royal Society of Antiquaries in May 1814, and then he took a copy of the Rosetta Stone inscription away with him from Welbeck Street to the relative tranquility of Worthing and spent the summer and fall studying Egyptian, when he was not attending to his medical patients.

Apart from his exceptional scientific mind and his broad knowledge of languages, Young brought to the problem one other extremely valuable and relatively uncommon ability. He had trained himself to sift, compare, contrast, retain and reject large amounts of visual linguistic data in his mind. This ability has been a *sine qua non* for serious decipherers ever since Young and Champollion, as I have described in my two books on decipherment: *Lost Languages: The Enigma of the World's Undeciphered Scripts* and *The Man Who Deciphered Linear B: The Story of Michael Ventris.* (Although outsiders to decipherment often like to imagine that in today's world computers could be programmed to accomplish such sifting, in reality the human factor remains all-important—mainly because only a human being can spot that two signs which objectively look somewhat different are in fact variants of the same sign. We can all learn, from our knowledge of a language, how to recognize the same phrase written in two very different kinds of handwriting; but the same task is extremely difficult for computers.)

In his teens and twenties, as we know, Young was celebrated for his penmanship in classical Greek, leading to the publication of *Calligraphia Graeca* with John Hodgkin. From this he developed a minutely detailed grasp of the Greek letter forms. Then, in his mid-thirties, he was called upon to restore some Greek and Latin texts written on heavily damaged papyri dug up in the ruins of Herculaneum, the Roman town smothered along with Pompeii by the eruption of Mount Vesuvius in AD 79. The fused mass of papyri had first to be unrolled without utterly destroying them, and then interpreted by classical scholars capable of guessing the meaning of illegible words and missing fragments. The unrolling required Young's chemical skills (and those of Sir Humphry Davy); the interpretation demanded his forensic knowledge of classical languages. In neither activity was Young at all satisfied with his results, but his experience with the Herculaneum papyri made him keenly aware of the relevance of his copying skills to the arcane arts of restoring ancient manuscripts. As he noted in his biography of Porson, "those who have not been in the habit of correcting mutilated passages of manuscripts, can form no estimate of the immense advantage that is obtained by the complete sifting of every letter which the mind involuntarily performs, while the hand is occupied in tracing it".

The mass of unpublished Egyptian research manuscripts by Young, now kept at the British Library, bear out this claim. Much of his success in this field would be due to his indefatigable copying—often exquisitely and occasionally in color—of hieroglyphic and demotic inscriptions taken from different ancient manuscripts and carved inscriptions and also from different parts of the same inscription, followed by the word-by-word comparisons that such copying made possible. By placing groups of

Egyptian signs adjacent to each other, both on paper and in his memory, Young was in a position to see resemblances and patterns that would have gone unnoticed by other scholars. As his biographer Peacock wrote, after immersing himself in Young's manuscripts, "It is impossible to form a just estimate either of the vast extent to which Dr Young had carried his hieroglyphical investigations, or of the real progress which he had made in them, without an inspection of these manuscripts." They also serve as a reminder, if one is still needed, of how unfounded are some modern claims that Young was a dilettante scholar.

It was his powerful visual analysis of the hieroglyphic and demotic inscriptions on the Rosetta Stone that gave Young the inkling of a crucial discovery. He noted a "striking resemblance", not spotted by any previous scholar, between some demotic signs and what he called "the corresponding hieroglyphics"—the first intimation that demotic might relate directly to hieroglyphic, and not be a completely different script, somewhat as a modern cursive handwritten script partly resembles its printed equivalent. We can see this relationship from the drawing he published showing the last line of the Rosetta inscription in hieroglyphic (which includes a cartouche), demotic and Greek, reproduced in Figure 10.3. If you examine the hieroglyphic and the demotic signs, you can see that some resemble each other. Equally clear, however, is that other "corresponding" signs do not.

The clinching evidence for the truth of this partial resemblance came with the publication of several manuscripts on papyrus in the monumental French survey mentioned earlier, *Description de l'Égypte*, the most recent volume of which Young was able to borrow in 1815. He later wrote:

I discovered, at length, that several of the manuscripts on papyrus, which had been carefully published in that work, exhibited very frequently the same text in different forms, deviating more or less from the perfect resemblance of the objects intended to be delineated, till they became, in many cases, mere lines and curves, and dashes and flourishes; but still answering, character for character, to the hieroglyphical or hieratic writing of the same chapters, found in other manuscripts, and of which the identity was sufficiently indicated, besides this coincidence, by the similarity of the larger tablets or pictural representations, at the head of each chapter or column, which are almost universally found on manuscripts of a mythological nature.

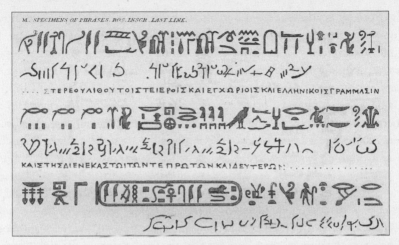

Figure 10.3 *Specimens of phrases from the last line of the Rosetta Stone inscription in hieroglyphic/demotic/Greek scripts, as shown in Young's* Encyclopaedia Britannica *article, "Egypt", published in 1819.*

In other words, Young was able to trace how the recognizably pictographic hieroglyphs, showing human figures, animals, plants and objects of many kinds, had developed into their cursive equivalents in the hieratic and demotic scripts.

But if the hieroglyphic and demotic scripts resembled each other visually in many respects, did this also mean that they operated on the same *linguistic* principles? If so, it posed a major problem, because the hieroglyphic script was generally supposed to be purely conceptual or symbolic (except for the foreign names in the cartouches, as suggested by de Sacy), whereas the demotic script was supposed (by Åkerblad) to be purely alphabetical. The two views could not be satisfactorily reconciled, if some of the signs in the demotic scripts were in fact hieroglyphic in origin.

So Young took the next logical step and made another important discovery. He told de Sacy in a letter in August 1815: "I am not surprised that, when you consider the general appearance of the [demotic] inscription, you are inclined to despair of the possibility of discovering an alphabet capable of enabling us to decipher it; and if you wish to know my 'secret', it is simply this, that no such alphabet ever existed". His conclusion was that the demotic script consisted of "imitations of the hieroglyphics ... mixed with letters of the alphabet." It was neither a purely conceptual or symbolic script, nor an alphabet, but a mixture of the two. As Young wrote a little later, employing an analogy for the demotic script that perhaps only a polymath such as he could have come up with, "it seemed natural to suppose, that alphabetical characters might be interspersed with hieroglyphics, in the same way that the astronomers and chemists of modern times have often employed arbitrary marks, as compendious expressions of the objects which were most frequently to be mentioned in their respective sciences." A modern, non-scientific example of the same idea would be such 'compendious' signs as $, £, %, =, +, which represent concepts non-phonetically, and often appear adjacent to alphabetic letters.

Young was correct in these two discoveries about the relationship between the hieroglyphic and demotic scripts. But we must also note that the discoveries did not now lead him to make a third discovery. He did not question the almost-sacred notion that the *hieroglyphic* script was purely symbolic. He continued to adhere to the view that the only phonetic elements in the hieroglyphic script were to be found in the foreign names in the cartouches, as first suggested by de Sacy. The idea that the hieroglyphic script as a whole might be a mixed script like the demotic script was to be the revolutionary breakthrough of Champollion.

As yet, in 1815, Young and Champollion, who started work on the Rosetta Stone around the same time as Young, had had little contact with each other. The previous November, Champollion had written to the president of the Royal Society from his base in Grenoble, enclosing his new book on Egypt and requesting some clarifications of parts of the Rosetta inscription which were not clear in his French copy; and Young, as the society's foreign secretary, had willingly obliged him, while also adding that Champollion might wish to consult his own conjectural translation of the Rosetta Stone which he had recently sent to de Sacy, one of Champollion's teachers when he had studied in Paris. This was really all that had passed between Young and Champollion. Young must therefore have been very taken aback to receive a letter from de Sacy written from Paris in July which openly warned him about his ex-student:

> If I might venture to advise you, I would recommend you not to be too communicative of your discoveries to M. Champollion. It may happen that he may hereafter make pretension to the priority.

He seeks, in many parts of his book, to make it believed that he has discovered many words of the Egyptian inscription of the Rosetta Stone: but I am afraid that this is mere charlatanism: I may add that I have very good reasons for thinking so.

De Sacy was writing a mere month after the battle of Waterloo, in which Champollion had supported the defeated Napoleon, and it was clear from other remarks de Sacy made that the revolutionary ex-student and the royalist former professor were now deeply divided by politics in addition to linguistic scholarship. Even so, de Sacy's explicit warning had a prophetic ring, and to some extent put Young on his guard against his leading competitor. "Since Champollion was obsessed by anything to do with ancient Egypt, this warning was partly justified, and should not be dismissed merely as academic jealousy on the part of a former teacher," writes the Egyptologist John Ray. What had started as a frank exchange of letters between Champollion and Young in 1814-15 would never develop into a major intellectual correspondence—in stark contradiction to Young's exchanges about the wave theory of light with French physicists during the same period, which we shall come to in the next chapter. In 1815, Young already felt he was the leader in the hieroglyphic field, but Champollion, though considerably younger, was not willing then, or at any point thereafter, to accept a subservient role.

Over the next three years, Young made a number of solid contributions to the decipherment of hieroglyphic and demotic. For example, he identified hieroglyphic plural markers and various numerical notations and a special sign used to mark feminine names. But his most important further discovery, following his two insights into the demotic-hieroglyphic relationship, arose from Barthélemy's idea that the cartouches expressed royal or religious names and de Sacy's idea that foreign names in the cartouches might be spelt phonetically.

There were six cartouches on the Rosetta Stone. From the Greek translation, these cartouches clearly had to contain the name Ptolemy (Ptolemaios, in Greek). Three of them looked like this:

and the other three like this:

Young postulated that the longer cartouche wrote the name of Ptolemy with a title, as suggested by equivalents in the Greek inscription, which read "Ptolemy, living for ever, beloved of Ptah".

This enabled Young to match the hieroglyphic signs in the short cartouche with known letters and phonetic values. Here is what he deduced, along with today's accepted phonetic value:

Hieroglyph	Young value	Today's value
□	P	P
⌒	T	T
𓏏𓍯	"not essentially necessary"	O
🐆	LO or OLE	L
⊂	MA or simply M	M
𓏭𓏭	I	I or Y
∏	OSH or OS	S

And here is his reasoning for these values:

The square block and the semicircle answer invariably in all the manuscripts to characters resembling the P and T of Åkerblad, which are found at the beginning of the enchorial name [i.e., the assumed name of Ptolemy written in demotic]. The next character, which seems to be a kind of knot, is not essentially necessary, being often omitted in the sacred characters [i.e., hieroglyphic], and always in the enchorial. The lion corresponds to the LO of Åkerblad; a lion being always expressed by a similar character in the manuscripts; an oblique line crossed standing for the body, and an erect line for the tail: this was probably read not LO but OLE; although, in more modern Coptic,

OILI is translated as ram; we have also EIUL, a stag; and the figure of the stag becomes, in the running hand [i.e., demotic or hieratic], something like this of the lion. The next character is known to have some reference to "place", in Coptic MA; and it seems to have been read either MA, or simply M; and this character is always expressed in the running hand by the M of Åkerblad's alphabet. The two feathers, whatever their natural meaning may have been, answer to the three parallel lines of the enchorial text, and they seem in more than one instance to have been read I or E; the bent line probably signified great, and was read OSH or OS; for the Coptic SHEI seems to have been nearly equivalent to the Greek sigma. Putting all these elements together we have precisely PTOLEMAIOS, the Greek name; or perhaps PTOLEMEOS, as it would more naturally be called in Coptic.

This passage was worth quoting at length in order to show that Young was sometimes capable of poor reasoning as well as his typical acuity. His analysis of Ptolemy's cartouche was mostly on target, but he was plainly wrong about the value of the knot, and also wrong in assuming that some of the phonetic values might be syllabic rather than alphabetic. He was less successful with the cartouche of a Ptolemaic queen, Berenice, which he guessed to be hers from a copy of an inscription beside her portrait in the temple complex of Karnak at Thebes. With the two cartouches taken together, Young was able to assign six phonetic values correctly, three partly so, while four were assigned incorrectly: the beginnings of his hieroglyphic 'alphabet'.

In 1818, Young summarized his Egyptian labors in a magnificent article, "Egypt", in the supplement to the fourth edition of the *Encyclopaedia Britannica*, which appeared in 1819. Here he published a vocabulary in English offering equivalents for 218 demotic and 200 hieroglyphic words, including proper names, things and numerals, a portion of which is shown in Figure 10.4; his phonetic values for thirteen hieroglyphs, cautiously headed "Sounds?"; and a "Supposed enchorial alphabet" for the demotic script. About 80 of his demotic-hieroglyphic equivalents have stood the test of time until today—an impressive record. Nothing remotely resembling this article had been published before on the subject of ancient Egyptian writing. Despite the fact that Young's results were "mixed up with many false conclusions," noted Francis Llewellyn Griffith, a highly respected Egyptologist working a century or so after him, "the method pursued was infallibly leading to definite decipherment." Young's article on "Egypt" was a landmark. But it was also anonymous: not until 1823 did he publish on Egypt under his own name. This self-effacement would undoubtedly encourage Champollion in his desire to avoid giving Young more credit than he absolutely had to.

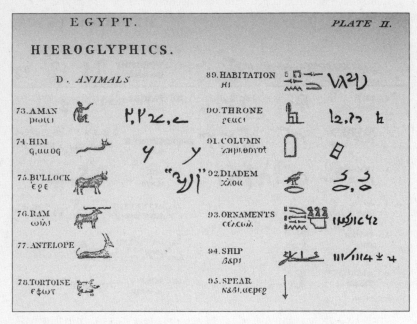

Figure 10.4   *Part of Young's hieroglyphic and demotic vocabulary, as shown in his*
Encyclopaedia Britannica *article, "Egypt", published in 1819.*

Of course most of those interested in ancient Egyptian inscriptions— a growing number of travelers and scholars curious and wealthy enough to visit Egypt and collect its antiquities—knew perfectly well who the anonymous author of the *Encyclopaedia Britannica* article was. In the 1820s, Young (like Champollion in France) was in constant communication with many such people, in a determined effort to obtain as many copies of new inscriptions and manuscripts as he could. One of them, the flamboyant circus actor, engineer, explorer and excavator Giovanni Belzoni (who first tunneled under the middle pyramid at Giza), wrote in his grand folio *Belzoni's Travels*, published in 1820:

> I have the satisfaction of announcing to the reader, that, according to Dr Young's late discovery of a great number of hieroglyphics, he found the names of Nichao and Psammethis his son, inserted in the drawings I have taken of this tomb. It is the first time that hieroglyphics have been explained with such accuracy, which proves the doctor's system beyond doubt to be the right key for reading this unknown language; and it is to be hoped, that he will succeed in completing his arduous and difficult undertaking, as it would give to the world the history of one of the most primitive nations, of which we are now totally ignorant.

Champollion was not even mentioned by Belzoni. In 1820, the French scholar was still virtually lost in the maze of the hieroglyphs. In Chapter 15, "Dueling with Champollion," we shall see how Champollion extricated himself in 1821-22, and in a few short months overtook Young. But before coming to that, we must return to Young's interests in natural philosophy and see how his undulatory theory of light—still very controversial when we left it in 1807—was at long last put on a firm scientific footing.

Chapter 11

Waves of Enlightenment

"I dare say poor Fresnel, if he had lived, would have preferred his share of the honor as much as I do mine. It was before I knew you that mine was earned; and acute suggestion was then, and indeed always, more in the line of my ambition than experimental illustration."

Young, letter to his sister-in-law Emily Earle, 1827

Among the long-known optical phenomena yet to be convincingly explained by the wave theory of light in the 1810s was something called double refraction. Anyone who has seen the effect of a colorless, transparent crystal of Iceland spar—a form of calcite so called because it was first produced in Iceland in the seventeenth century—will know this phenomenon. If you place a thin slice of Iceland spar on top of a printed page and orient it in a particular way, you see a double image of the printed letters underneath the crystal. Iceland spar bends an incident ray of light into *two* refracted rays, not just the usual single refracted ray observed when light is incident on glass or water (see Figure 11.1).

One of these rays is known as the *ordinary* (or regular) ray. It behaves like the usual refracted ray, with its angle of incidence and its angle of refraction related to each other by Snell's law (as mentioned in Chapter 7, "Let There Be Light Waves"). The other ray is known as the *extraordinary* (or irregular) ray. Its refraction is not governed by Snell's law. This is because the nature of the ordinary ray and the extraordinary ray differs in one fundamental respect, as we shall now come to understand.

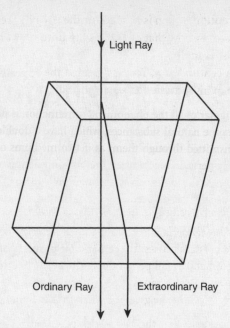

Light Ray

Ordinary Ray   Extraordinary Ray

Figure 11.1 *Double refraction of a light ray incident at right angles to one face of a calcite crystal. It was first observed with Iceland spar.*

Double refraction was first described in detail by Erasmus Bartholin in 1670. Soon after, Christiaan Huygens performed his own experiments with Iceland spar and observed an additional peculiarity. When Huygens used two identical, flat polished plates of the special crystal to view a small object, he saw the object either doubled or quadrupled, depending on the relative position of the two plates. It seemed that in some positions, the lower plate had the power to split the two rays from the upper plate into four rays, but not in other positions. When Huygens rotated the lower plate, keeping the upper plate still, the double image and the quadruple image alternated: he saw two images, then four images, then two images, then four images, and so on, as he kept turning the plate. Why should the lower plate of Iceland spar apparently lose its power to split light rays and produce only a double image in certain positions? Unable to theorize this, Huygens hazarded the comment: "It seems as though, as it passes the upper plate, the [regular] ray has lost something that is necessary to bring the matter into motion, which is needed for the irregular refraction [in the lower plate]. But to say how that operates—up until now I have discovered nothing that satisfies me."

More than a century later, Young repeated the experiment and wrote about the puzzle in his *Natural Philosophy* in 1807:

> The most singular of all the phenomena of refraction is perhaps the property of some natural substances, which have a double effect on the light transmitted through them, as if two mediums of different densities freely pervaded each other, the one only acting on some of the rays of light, the other on the remaining portion. These substances are usually crystallized stones, and their refractions have sometimes no further peculiarity; but the rhomboidal crystals of calcarious spar [the name calcite was not coined until 1845], commonly called Iceland crystals, possess the remarkable property of separating such pencils of light, as fall perpendicularly on them, into two parts, one of them only being transmitted in the usual manner, the other being deflected towards the greater angle of the crystal. ...
>
> It is also remarkable, that the two portions of light, thus separated, will not be further subdivided by a transmission through a second piece [of Iceland spar], provided that this piece be in a position parallel to that of the first; but if it be placed in a transverse direction, each of the two pencils will be divided into two others; a circumstance which appears to be the most unintelligible of any that has been discovered respecting the phenomena of double refraction.

Experiments with double refraction using various kinds of crystal done by others around this time were still more confusing. If light was shone on the colored mineral tourmaline, the ordinary ray was seen to be absorbed after passing through a small thickness of the crystal, while the extraordinary ray was transmitted by the crystal. When two sections from one crystal of tourmaline were put one behind the other in exactly the same orientation as they had before they were cut, light passed through both of them. But if the second section of tourmaline was now turned through 90 degrees, no light passed: the extraordinary ray was stopped. However, when the second slice was turned through a further 90 degrees, the extraordinary ray again passed through it.

If the second crystal of Iceland spar in Young's experiment was replaced by a thin slice of tourmaline (thin enough for the ordinary ray not to be totally absorbed), a further intriguing effect was observed. With the slice in one particular position, it transmitted only the ordinary ray from the Iceland spar and cut off the extraordinary ray. But when the slice was turned through 90 degrees, the situation was reversed and the extraordinary ray was transmitted while the ordinary ray was cut off. At intermediate angles, the tourmaline transmitted part of the ordinary ray and part of the extraordinary ray.

In our own time, a similar effect is well known with so-called Polaroid sunglasses, which are designed to reduce glare. If you look at a sunny scene through two pairs of Polaroid sunglasses held in the same orientation, light passes through both sets of lenses. But if you turn one pair of sunglasses through 90 degrees, keeping the two lenses for the right (or the left) eye superimposed, the sunlight is almost entirely cut off before it can reach the eye. At an intermediate angle, part of the sunlight remains visible.

From these observations, it is clear that light has some kind of orientation in addition to its direction of propagation, and that different light rays can have different orientations. Today, the extraordinary ray in double refraction is said to be differently *polarized* from the ordinary ray. Polaroid sunglasses work by eliminating as much horizontally polarized light as possible, since glare consists largely of horizontally polarized light, and transmitting vertically polarized light.

In the seventeenth century, Huygens and Newton did not refer to 'polarization' as such—the term was introduced only in 1809 by Etienne Malus—but they accepted as a fact the orientability of light. Their problem, and the problem of their early nineteenth-century followers, was how to explain this fact theoretically. How could a light corpuscle—as imagined by Newton or Pierre-Simon Laplace (a convinced corpuscularist)—or a light wave—the concept preferred by Huygens or Young—be physically oriented, other than in the direction of propagation of a light ray?

Newton's *Opticks* devotes a number of pages to a discussion of double refraction, but Newton plainly felt frustrated by it. Query 26 in this section of his great work reads: "Have not the rays of light several sides, endued with several original properties?" The idea of "sides" for light was Newton's stab at the future concept of polarization. Query 27 sums up: "Are not all hypotheses erroneous which have hitherto been invented for explaining the phenomena of light, by new modifications of the rays? For those phenomena depend not upon new modifications, as has been supposed, but upon the original and unchangeable properties of the rays." The only plausible hypothesis that corpuscularists could propose to account for double refraction was that corpuscles were sorted into two rays—ordinary and extraordinary—by the structure of a crystal, according to a corpuscle's shape; and that a second crystal then filtered them again according to their shape, blocking some and letting through others. "It would be like trying to put square pegs into square holes; they only fit if oriented certain ways," writes the physicist Arthur Zajonc. But nobody in this early period had any credible notion as to why the various corpuscles in a given light ray should vary in shape (in itself unlikely) and what this might entail in mechanical terms. If the corpuscles really were similar to bullets from a gun, there seemed to be no compelling reason why they should not all be identical in shape and oriented in only one direction.

Huygens, by contrast, gave the wave theory a small advantage over the corpuscular theory. He came up with a mathematical model that enabled him to calculate, for a given substance, the direction of the extraordinary ray; and his results agreed with experiment. His model assumed that the ether inside the double-refracting crystal propagated two light waves—one spherical (the ordinary ray), the other ellipsoidal (the extraordinary ray); the latter was ellipsoidal because its velocity was different in different directions. This assumption seemed reasonable given the fact that the various kinds of crystals exhibiting double refraction were known to have other properties, too, such as thermal conductivity and mechanical elasticity, that varied with the orientation of the crystal. (Tourmaline is today used in piezoelectric devices, because, along particular axes, a tourmaline crystal generates an electric charge when put under mechanical stress and also changes its shape when a voltage is applied to it.) But Huygens's wave model was completely unable, like the corpuscular model, to account for the strange behavior of polarized light when it was incident on a second double-refracting crystal.

Thus in 1807, when Young published his *Natural Philosophy*, neither theory—corpuscular or undulatory—gave much of an account of polarization, or for that matter a fully satisfactory account of reflection, refraction and diffraction, as discussed in Chapter 7. The only experiments that seemed to favor the wave theory unequivocally were Young's experiments on interference, but, as presented in *Natural Philosophy*, they lacked the kind of precision and mathematical rigor increasingly expected by physicists. In bald truth, Young's theory of interference made so little initial impression that there is hardly a single allusion to the interference of light in any work on optics published in Britain or abroad between the time of Henry Brougham's attack in 1802-04 and the more mathematical, rigorous work of Augustin Fresnel and Dominique Arago in 1816; when the astronomer William Herschel (father of John) communicated three papers on Newton's rings to the Royal Society in 1807, 1809 and 1810, Herschel did not even mention Young's interference explanation of the colors of the rings published by the Royal Society in 1802.

During the next decade or so, however, from 1808, there was an explosion of new experimental evidence in optics and attempts at theoretical explanation, as the debate hotted up between corpuscularists and undulationists. Much of the experiment and argument focused on polarization, and in the 1820s, it would be the undulatory explanation of this particular phenomenon that would convert all but the most fanatical of corpuscularists (they of course included Young's detractor, Brougham). George Peacock, who was a young Cambridge mathematician in this period of flux and ferment, catches the excitement among physicists generated by these developments—most of which came out of France—in his biography of Young:

> In the intermediate period Laplace [a corpuscularist] published his celebrated memoir on the double refraction of Iceland spar [in 1809]: Malus [an undulationist] had discovered the polarization of light by reflection, and was engaged in a brilliant series of researches connecting his discovery with the optical properties of crystalline bodies, when a premature death brought his labors to a close: [David] Brewster [a corpuscularist] was enriching every department of experimental optics with the most remarkable speculations and discoveries: Arago [an undulationist] had found out the colors of crystalline

plates produced by polarized light, and though less fertile than some of his contemporaries in the number of his contributions to the science, he was second to none of them in the critical sagacity with which he analyzed their labors: [Jean-Baptiste] Biot [a corpuscularist] was combining theoretical and practical researches with a success and ingenuity which seemed to promise him the first place amongst optical discoverers, when it was his misfortune to waste his energies and compromise his reputation in the proposition and obstinate maintenance of his theory of moveable polarization: at a later period [from 1815 onwards], the labors of Fresnel [an undulationist], who—though treading generally in the footsteps of Young, required no foreign aid either to guide or support him—were destined to give unity and system to the vast mass of facts and theories which his predecessors had accumulated and prepared.

Young himself, in this period, was trying to establish his credentials as a physician, rather than as a physicist. He did few optical experiments after 1807, but he was keenly interested in the optical works of others and wrote several anonymous reviews of them and a major article, "Chromatics", for the *Encyclopaedia Britannica*. Although he never wavered in his belief in the undulatory theory, its intractable experimental difficulties disturbed him.

Reviewing Malus's work on polarization in 1810, for instance, Young commented somewhat pessimistically, despite the fact that Malus himself did not draw the same conclusion as he did· "This statement appears to us to be conclusive with respect to the insufficiency of the undulatory theory, in its present state, for explaining all the phenomena of light. But we are not therefore by any means persuaded of the perfect sufficiency of the projectile [corpuscular] system … [Much] more evidence is still wanting before [the question] can be positively decided."

Five years later, in a letter to Brewster in 1815, Young stated: "With respect to my own fundamental hypotheses respecting the nature of light, I become less and less fond of dwelling on them, as I learn more and more facts like those which Mr Malus discovered: because, although they may not be incompatible with these facts, they certainly give us no assistance in explaining them."

And in his article in the *Encyclopaedia Britannica*, written in late 1817 in the heat of the debate, Young began:

> But notwithstanding all that has hitherto been done, it appears to be utterly impracticable, in the present state of our knowledge, to obtain a satisfactory explanation of all the phenomena of optics, considered as mechanical operations, upon any hypothesis respecting the nature of light that has hitherto been advanced: it will therefore be desirable to consider the facts which have been discovered, with as little reference as possible to any general theory …

Nonetheless, he put forward the undulatory theory—of course anonymously—as the only real hope for progress:

> [A]t the same time, it will be absolutely necessary, as a temporary expedient, to borrow from the undulatory system Dr Young's law of the interference of light, as affording the only practicable mode of connecting an immense variety of facts with each other, and of enabling the memory to retain them; and this adoption will be the more unexceptionable, as many of the most strenuous advocates for the projectile [corpuscular] theory have been disposed, especially since the experiments of Mr Arago and Mr Fresnel, to admit the truth of the results of all the calculations, in which this law has been employed.

Reviewing Laplace's highly mathematical, corpuscular explanation of double refraction, with which he pungently disagreed, Young made a significant suggestion. The acoustician Ernst Chladni had observed that sound traveling through wood—a rod of Scotch fir in particular—had a slightly higher velocity in one direction, along the grain, than in another direction, at an angle to the grain, due to the wood's being more elastic in the first direction than in the second. (The ratio of the different velocities is about five to four.) Perhaps, wrote Young, in double refraction there is a similar kind of difference in the "structure of the elementary atoms of the crystal" as in the grain of the wood? This would produce the ellipsoidal wave motion suggested by Huygens, and account for the extraordinary ray. But acute as Young's analogy was, it suffered from the crucial disadvantage that it did not account for the existence of the ordinary ray.

It also shows how he was misled by his persistent comparisons of light and sound. Light and sound could both be reflected, refracted and diffracted. They could also interfere with themselves, as shown by the double-slit experiment and by the phenomenon of beats (described in Chapter 7). Sound was long established to be an alternating compression and rarefaction of the air (a true vacuum is, of course, soundless)—a pressure wave that transmitted itself *longitudinally*, that is, along the path of the sound's propagation. By analogy, light was therefore thought by Young to be a longitudinal compression and rarefaction of the ether, in the direction of the light ray. But if this assumption were correct, it would provide no explanation for the polarization of light. Sound does not exhibit any polarized phenomena; there is, for example, no double refraction of sound, with an 'ordinary' sound wave and an 'extraordinary' sound wave. For sound, being a longitudinal wave, by definition, can have only *one* intrinsic orientation, along the line of propagation.

For a long time, the analogy with sound "blinded" Young and others, including Arago, to the "secret of polarization", in the words of his biographer Alex Wood. In "Chromatics", Young writes: "It is certainly easier to conceive a detached particle, however minute, distinguished by its different sides, and having a particular axis turned in a particular direction, than to imagine how an undulation, resembling the motion of the air which constitutes sound, can have any different properties, with respect to the different planes which diverge from its path." In other words, profoundly unsatisfactory though the corpuscular theory of polarization was, Young found it impossible to explain polarization on the undulatory theory, because he imagined that light waves in the ether must be transmitted longitudinally, like sound waves in air, along the path of a light ray.

Then, in January 1817, a long-frustrated Young finally speculated on a possible theoretical solution to the polarization problem. Writing to Arago, he stated:

> I have ... been reflecting on the possibility of giving an imperfect explanation of the affection [refraction] of light which constitutes polarization, without departing from the genuine doctrine of undulations. It is a principle in this theory, that all undulations are simply propagated through homogenous mediums in concentric spherical

surfaces like the undulations of sound, consisting simply in the direct and retrograde motions of the particles in the direction of the radius, with the concomitant condensation and rarefactions. And yet it is possible to explain in this theory a transverse vibration, propagated also in the direction of the radius, and with equal velocity, the motions of the particles being in a certain constant direction with respect to that radius; and this is a *polarization*.

The key word here is *transverse*; we should not bother too much about the precise wording of the rest of Young's explanation (which again drags in an analogy with sound). A water wave is a transverse vibration; a sound wave is not—it is longitudinal. The water moves up and down transversely, that is, at right angles to the direction of propagation of the wave; the air molecules move longitudinally, in the same direction as the sound propagates. Young was making the radical proposal to Arago that light might be a longitudinal wave with a transverse component.

Polarization would be explicable, at least in principle, with a transverse wave. A transverse wave vibrates in a plane perpendicular to the axis of propagation; however, within this perpendicular plane, the vibration can take any orientation at all with respect to the axis. It can, for instance, be vertically polarized, like a water wave, and not vibrate at all at any other angle to the axis, including horizontally. Or it can be horizontally polarized, and not vibrate at all at any other angle to the axis, including vertically. We can imagine these two situations with an elastic cord. You fix one end of the cord to a hook on a wall, and then you make waves in the cord. If you move your hand rhythmically up and down—that is, vertically— you can create a transverse wave in the cord that is vertically polarized. If instead you move your hand from side to side, horizontally, you get a horizontally polarized transverse wave (if you ignore the inevitable sagging of the cord). Then imagine threading the cord through a slit in a card located between you and the hook. If the slit is held vertical, the vertically polarized wave will pass through the slit, whereas the horizontally polarized wave will be stopped by it. Turn the slit through 90 degrees so that it is horizontal, and the horizontally polarized wave will be transmitted, whereas the vertically polarized wave will be cut off. The crystal structure of Iceland spar or tourmaline could act in a similar way to the slit, cutting off the extraordinary ray in one crystal orientation and the ordinary ray in a second crystal orientation at 90 degrees to the first one—assuming, that is, the light in the ordinary ray is polarized in a direction at 90 degrees to the light in the extraordinary ray. Hence, the tourmaline slice's ability to cut off the ordinary ray but transmit the extraordinary ray coming from a piece of Iceland spar in one orientation of the slice, and transmit the ordinary ray but cut off the extraordinary ray when the slice is turned through 90 degrees.

Today, following the pioneering work of James Clerk Maxwell in the second half of the nineteenth century, we know that light is an entirely transverse *electromagnetic* wave. An electric field and a magnetic field vibrate in a plane perpendicular to the direction of propagation, with the two fields always at right angles to each other (Figure 11.2). Polarization of light means confining the electric field (and hence the magnetic field) to a certain orientation. In *plane-polarized* light—as transmitted by Polaroid sunglasses—the electric field vibrates in only one direction within the plane.

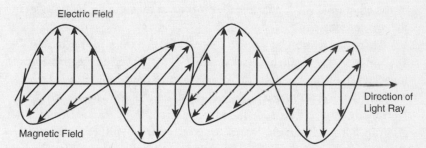

Figure 11.2   *Light is an electromagnetic wave, in which the electric and magnetic fields vibrate at right angles to each other and also at right angles to the direction of propagation of the wave, that is, transversely, not longitudinally.*

Young had provided the seed concept to explain polarization—the transverse wave—but it would not be true to say that he nurtured this into a full-blown theory. As so often with him, he laid the foundation, but left others to consolidate the theory. Fresnel, advised by Arago, had developed Young's extensive early work on interference and diffraction into a convincing system from 1815 onwards. Now he did the same for Young's *aperçu* that light could be transverse; indeed, Fresnel transformed it into a model that really worked and could rationalize many of the confusing phenomena of polarization. In 1817, Young had proposed a small transverse component to light, but in the same breath he had retained a far larger longitudinal component. Fresnel, by 1821, was able to show via mathematical methods that polarization could be explained only if light was *entirely* transverse, with no longitudinal vibration at all.

The mechanically inclined Young was not convinced. In 1823, he wrote in the *Encyclopaedia Britannica*: "This hypothesis of Mr Fresnel is at least very ingenious, and may lead us to some satisfactory computations: but it is attended by one circumstance which is perfectly *appalling* in its consequences." Only in solids, Young emphasized, had transverse waves with "lateral" (i.e., horizontal) vibrations been observed. Fresnel's hypothesis would mean that "the luminiferous ether, pervading all space, and penetrating almost all substances, is not only highly elastic, but absolutely solid!!!" Young was not much of a user of exclamation marks, and his point was certainly a penetrating one, which Fresnel's theory could only ignore. Young's criticism was not fully answered until the internal contradictions of the ether finally led to its abandonment as a medium for light waves in the time of Einstein.

There was tension between Young and Fresnel from the beginning. In 1815, Fresnel had begun his work on light by rediscovering the principle of interference, being unaware of Young's publications of 1800-07. He was working far from the scientific circles of Paris in his home village in Normandy using a few crude instruments made by a local blacksmith; he also did not read English. It was Arago who drew Fresnel's attention to Young's work, published in the Royal Society's *Philosophical Transactions* and in Young's *Natural Philosophy*. Fresnel acquiesced gracefully—though in truth he had little choice, given the clear-cut published evidence— telling Young in 1816 that "if anything could console me for not having the advantage of priority, it would be having been brought into contact with a scholar who has enriched physics with so many important discoveries, and that has contributed not a little to increase my confidence in the theory which I adopted." Young was generous in his response to Fresnel, though at times proprietorial. In the 1820s, as Fresnel made important original progress, unaided by Young, the contact between them was not always smooth, but it never degenerated into anything like the suspicion and open rivalry that would mark Young's relationship with Fresnel's fellow countryman Jean-François Champollion (who was almost the same age as Fresnel) during exactly the same period. Indeed, Young had a hand in the award of the distinguished Rumford medal of the Royal Society to Fresnel, just before his premature death from tuberculosis in 1827, at the age of only 39. (As foreign secretary of the society, Young sent Fresnel the medal along with an official letter of congratulation.)

An important reason for their better relationship, as compared with Young and Champollion, was that Young and Arago became close friends and Arago acted as a trusted interpreter between Young and Fresnel. After Young's death, an affectionate Arago wrote a memoir of Young in which he vividly recalled this aspect of his role at his first meeting with Young in Worthing:

> In the year 1816, I passed over to England with my learned friend M. Gay-Lussac. Fresnel had then just entered in the most brilliant manner into the career of science by publishing his *Memoire sur la Diffraction*. This work, which, according to us, contained a *capital* experiment, irreconcilable with the Newtonian theory of light, naturally became the first object of our communication with Dr Young. We were astonished at the numerous restrictions he put upon our commendations, and in the end he told us that the experiment about which we made so much ado was published in his work on *Natural Philosophy* as early as 1807. This assertion did not appear to us correct, and this rendered the discussion long and minute. Mrs Young was present, and did not appear to take any interest in the conversation; but, as we knew that the fear, however puerile, of passing for learned ladies— of being designated *blue-stockings*—made the English ladies very reserved in the presence of strangers, our want of politeness did not strike us till the moment Mrs Young rose up suddenly and left the room. We immediately offered our most urgent apologies to her husband, when Mrs Young returned, with an enormous quarto under her arm. It was the first volume of the *Natural Philosophy*. She placed it on the table, opened it without saying a word at [a certain page], and pointed with her finger to a figure where the curved line of the diffracted bands, on which the discussion turned, appeared theoretically established.

By 1827, the year of Fresnel's death, the Young-Fresnel undulatory theory of light, as it would now become known, was sufficiently established to explain all the major phenomena of light—reflection, refraction, diffraction and polarization—in a quantitative way that left little room for doubt that it was superior to the corpuscular theory. Even the magisterial corpuscularist Laplace had been won over. In Britain, the theory received a sort of official blessing in an encyclopedia article written by Sir John Herschel. He wrote:

> The unpursued speculations of Newton [about undulations in light], and the opinions of Hooke, however distinct, must not be put in competition, and, indeed, ought scarcely to be mentioned, with the elegant, simple, and comprehensive theory of Young—a theory which, if not founded in nature, is certainly one of the happiest fictions that the genius of man ever invented to grasp together natural phenomena, which, at their first discovery, seemed in irreconcilable opposition to it. It is, in fact, in all its applications and details, one succession of *felicities*; insomuch, that we may be almost induced to say, if it be not true, it deserves to be so.

Then Herschel paid tribute to Fresnel, too:

> [W]e must not, in our regard for one great name, forget the justice which is due to the other; and to separate them and assign to each his share would be as impracticable as invidious, so intimately are they blended together throughout every part of this system—early, acute and pregnant suggestion characterizing the one [Young], and maturity of thought, fullness of systematic development and decisive experimental illustration, equally distinguishing the other [Fresnel].

On this occasion, Young did not demur. He told his sister-in-law Emily that Herschel "has divided the prize very fairly" and remarked that the lately deceased Fresnel probably would have agreed. For, he admitted, his own ambition lay always more in the direction of "acute suggestion" than "experimental illustration". It was one of the perquisites of being a polymath—and also one of its penalties, as would become increasingly apparent to Young with age.

Chapter 12

Walking Encyclopedia

"The longer a person has lived the less he gains by reading, and the more likely he is to forget what he has read and learnt of old; and the only remedy that I know of is to write upon every subject that he wishes to understand, even if he burns what he has written."

Young, letter to Hudson Gurney, 1809

Nowhere in Young's works— not even in his *Natural Philosophy* of 1807—is his polymathy better displayed than in his contributions to the *Encyclopaedia Britannica*, written between 1816 and 1825. The unparalleled range of his subjects has already been mentioned in the introduction, and we have subsequently encountered remarkable examples in the shape of his contributions on "Egypt" and "Chromatics". In total, Young wrote 63 articles for a *Supplement* to the sixth edition of the *Britannica*, of which 46 were biographical. The most substantial and original of them were reprinted in his *Miscellaneous Works* in 1855; they comprise "Bridge", "Carpentry", "Chromatics", "Cohesion", "Egypt", "Herculaneum", "Languages", "Tides", "Weights and measures" and 23 biographies, ranging from men of science such as Joseph Lagrange, Etienne Malus and Count Rumford to Young's friend, the classical scholar Richard Porson. Young himself thought most highly of the three disparate articles, "Bridge", "Egypt" and "Tides", and of four of the biographies. Very few specialists nowadays would dare to attempt more than one, or at the most two, such major surveys for the *Encyclopaedia Britannica*, even with the help of a co-author.

The first edition of the encyclopedia, compiled "by a society of gentlemen in Scotland", appeared in 1771, two years before Young's birth. Its preface opens with a forthright declaration worthy of the eighteenth-century Enlightenment's hunger for scientific knowledge:

> Utility ought to be the principal intention of every publication. Wherever this intention does not plainly appear, neither the books nor their authors have the smallest claim to the approbation of mankind. ... Whoever has had occasion to consult Chambers, Owen, etc. or even the voluminous French *Encyclopédie* [of Denis Diderot], will have discovered the folly of attempting to communicate science under the various technical terms arranged in an alphabetical order. Such an attempt is repugnant to the very idea of science, which is a connected series of conclusions deduced from self-evident or previously discovered principles. It is well if a man be capable of comprehending the principles and relations of the different parts of science, when laid before him in one uninterrupted chain. But where is the man who can learn the principles of any science from a dictionary compiled upon the plan hitherto adopted? We will, however, venture to affirm, that any man of ordinary parts, may, if he chooses, learn the principles of agriculture, of astronomy, of botany, of chemistry, etc. etc. from the *Encyclopaedia Britannica*.

This was a founding prospectus destined to appeal to Young, and the editor of the new supplement, the energetic Macvey Napier of Edinburgh (who was later to be editor of the *Edinburgh Review*), was extremely keen to have Young as a contributor. Clearly Henry Brougham's notorious attack on Young a decade earlier had not in any way dampened Napier's editorial zeal. But when Young was first approached by him in mid-1814, (at the time he was studying the Rosetta Stone in Worthing), he refused the request to contribute. Professional success in medicine was still at the forefront of Young's ambitions. He would readily have agreed "under other circumstances", he said, "but I feel it a matter of necessity to abstain as much as possible from appearing before the public as an author in any department of science not immediately medical". As he told his sister-in-law Emily a couple of months later, "I have long been intending to write to you, but I am so much engaged in collecting materials for a new medical work that I have had no leisure to do anything else." He was referring to his treatise on consumptive diseases, which he forced himself to complete in only eight or nine months, and published in 1815. "It was not my own wish nor my intention to be so employed; but I am determined to make a last effort of this kind … which I think will insure me a certain degree of popularity in my profession".

Quite why Young recanted his decision in early 1816 is not clear, but presumably it was partly to do with the disappointing reception of this book and his sluggish medical practice. He may well have felt that he needed the income obtainable from 'popular' writing. But still he made anonymity a condition of his agreement when writing to Napier:

> I could not at present allow my name to be published as a contributor to the work; on the other hand, I could probably furnish you with some articles which you could scarcely obtain from other quarters: I should not refuse to do my best upon any subject of science, and I would consent to acknowledge all my contributions at the end of ten years from the present time. Knowing, however, the importance of names that are familiar to the public, I will not deny that you might fairly offer me a remuneration somewhat less liberal on this account; and I must therefore beg you to favor me with your sentiments on the subject as soon as is convenient to you.

Young accepted an offer of 16 guineas a sheet while his contributions remained anonymous, rising to 20 guineas if he went public; in the event, he agreed to put his name to his articles in 1823. These were very considerable sums, given that in 1820 Young parted with the copyright for the second edition of his *Introduction to Medical Literature* in exchange for a mere £100, which is about 95 guineas ("as I got nothing by the first edition, I was very ready to accept," he informed Gurney). In all, his *Encyclopaedia Britannica* articles covered about 380 quarto pages—"Languages" alone runs to some 33,000 words—which, at the preceding rates, should have earned their author between six and seven thousand pounds over the nine years he was a contributor.

Apart from the many biographies and the nine articles just mentioned, Young also agreed to take on the following subjects: annuities, bathing, fluents, hydraulics, life preservers, road-making, steam engines and the polarization of light (his translation of a French article by Arago, to which Young added some notes). Prodigious though this output was, editor Napier was not satisfied: it is clear from his correspondence preserved in the British Library that Napier "was continually trying to foist on to Young subjects which he was unwilling to tackle, or really felt himself incapable of tackling", writes Frank Oldham in Alex Wood's biography. "*Baths* I cannot refuse though I do not foresee anything very amusing in it," Young told Napier in 1816—after all, sea bathing was the *raison d'être* of Worthing as a health resort. "Craniology too I must accept though I am almost ashamed to be employed in such trash"—though in practice he evaded doing this article. But he later refused "Blasting and boring" on the grounds that, "For the last ten years I have paid no attention to the mechanical arts in any form—nor do I wish to renew my acquaintance with them—preferring general investigations to particular applications." In 1821, he turned down "Mining" and "Stone cutting". "There was a time in my life when I should have considered myself qualified to say something on mining"—recall that while a medical student in the 1790s, Young had toured Cornish mines with Gurney and descended deep mine shafts in the Harz Mountains of Germany—"but I have so totally changed my pursuits that I should be at a loss at present to know where to begin, and it would require the study of some years to enable me to write a tolerable article of a few pages on it." Of stone cutting, he said: "I never knew anything and have still less idea what is to be known than of mining." But in 1823, he yielded to a plea from Napier ("No subject comes amiss to you") and agreed to write a brief article on road-making: "I shall not even object to following Hannibal and Bonaparte in their road-making."

The biographies interested him less than the survey articles—with the exception of the one on Porson ("the only one I did *con amore*")—but, being Young, he did not skimp the necessary research. When writing about the mathematician Joseph-Louis Lagrange, he read more than a hundred of Lagrange's original papers—we know this because Young cites them, and sometimes summarizes them. He told Napier in 1820: "Lagrange will be an arduous task but I must not flinch from it; I cannot promise it till Christmas; it cannot but be long, probably longer than any of my biographical articles; but the labor will be much more than in proportion to its length." Writing to Gurney on the same subject, Young reflected profoundly on his general intellectual motivation. Part of what he said was quoted earlier in the chapter on his childhood, but it bears repeating in full and in its adult context:

> The biographical articles seldom *amuse* me much in writing; there is too little invention to occupy the mind sufficiently: I like a deep and difficult investigation when I happen to have made it easy to myself if not to all others—and there is a spirit of gambling in this, whether as by the cast of a die, a calculation *à perte de vue* [i.e., a farfetched calculation] shall bring out a beautiful and simple result, or shall be wholly thrown away. Scientific investigations are a sort of warfare, carried on in the closet or on the couch against all one's contemporaries and predecessors; I have often gained a signal victory when I have been half asleep, but more frequently found, on being thoroughly awake, that the enemy had still the advantage of me when I thought I had him fast in a corner—and all this, you see, keeps one alive.

In the comparable words of the inimitable Einstein: "I have little patience for scientists who take a board of wood, look for its thinnest part, and drill a great number of holes when the drilling is easy." Young, though less concise in his analogies, would have wholeheartedly agreed. Wide ranging though his interests were, he was never superficial or facile in his approach. A grateful Napier eventually hailed the newly public Young in his editor's preface, as a man "to whose profound and accurate knowledge, rare erudition, and other various attainments, this work is largely indebted in almost every department which it embraces."

Of all Young's writings for the *Encyclopaedia Britannica*, the article on "Egypt" is the most cited contribution today. Since it forms part of Chapter 10, "Reading the Rosetta Stone", and 15, "Dueling with Champollion", here we shall pass over it to what is probably his second most influential encyclopedia article. This is his survey, "Languages". Although not one of Young's own 'top three' contributions, unlike "Egypt", "Languages" occupies a secure niche in historical linguistics. It may be that Young slightly underrated it because it consisted almost entirely of two previously published articles by him on languages in the *Quarterly Review*, the first of which we have already mentioned: his 1813 review of Johann Christoph Adelung's *Mithridates*, which had triggered his interest in the Rosetta Stone.

"Languages" shows off Young's writing at its most felicitous, with cosmopolitan touches verging on the debonair. One feels that Young is drawing on a deep well of knowledge, fed by springs from his childhood, when he first became fascinated by the "Lord's Prayer" written in more than a hundred languages. "Of language in general this essay is not intended to treat, but merely of languages as they are distinct from each other," he writes.

> It is not, however, very easy to say what the definition ought to be that should constitute a separate language; but it seems most natural to call those languages distinct, of which the one cannot be understood by common persons in the habit of speaking the other, so that an interpreter would be required for communication between persons of the respective nations. Still, however, it may remain doubtful whether the Danes and the Swedes could not, in general, understand each other tolerably well, and whether the Scottish Highlanders and the Irish would be able to drink their whisky together, without an interpreter; nor is it possible to say, if the twenty ways of pronouncing the sounds, belonging to the Chinese characters, ought or ought not to be considered as so many languages or dialects, though they would render all oral intercourse between the persons so speaking the language actually impracticable. But, whether we call such variations different languages, or different dialects, or merely different pronunciations of the same dialect, it is obvious that they ought all to be noticed in a complete history of languages; and, at the same time, that the languages so nearly allied must stand next to each other in a systematical order; the perfection of which would be, to place the nearest together those languages, in which the number of coincidences in the signification of words, throughout the language, are the most numerous.

> It has sometimes been imagined, that all languages in existence present something like a trace of having been deduced from a common origin; and it would be difficult to confute this opinion by very positive evidence, unless every separate language had been very completely analyzed and examined by a person well acquainted with a variety of other languages, with which it might be compared. But

without such an examination, the opinion must remain conjectural only, and no more admissible as demonstrated, than the opinion of some empirics, that there is only one disease, and that the only remedy for it is brandy.

Having set out the challenge, he now proceeds to examine some four hundred languages—in considerable detail—and to group them into families, according to the degree of overlap between them: what he had called "the number of coincidences in the signification of words". It is a bravura performance, ranging through millennia and across continents from classical Greek, Coptic and Sanskrit to Chinese, Berber and Cherokee. But, careful scientist that he was, Young was keenly aware of the need for caution in seeking and identifying similarities between two words of the same meaning in two very different languages—a favorite sport of cranks. Ever since he had studied the limited number of sounds that the human voice was capable of producing, in his Göttingen University dissertation, Young had fully realized that many of the coincidences between words in different languages might simply be accidents. One coincidence on its own meant little or nothing. If, however, *several* words of the same meaning were similar, then the two languages might well be derived from a common source. Young estimated the odds in favor of this as ten to one with three words in common, rising to 1700 to one with six words in common, and 100,000 to one with eight shared words.

Such thinking and analysis led him to group together the Indian, the West Asiatic and almost all of the European languages. He wrote: "every one of them has too great a number of coincidences with some of the others, to be considered as merely accidental, and many of them in terms relating to objects of such a nature, that they must necessarily have been, in both of the languages compared, rather original than adoptive"—he meant fundamental terms like 'heaven', 'earth', 'day', 'father', 'mother', that native speakers are virtually certain to have invented rather than adopted from a foreign language. "The Sanskrit, which is confessedly the parent language of India, may easily be shown to be intimately connected with the Greek, the Latin, and the German, although it is a great exaggeration to assert anything like its complete identity with either of these languages."

Sir William Jones, the polymath mentioned in the introduction (whose Persian grammar Young had read as a teenager), had been the first scholar to make plain this still-surprising link between Sanskrit and Greek, in a well-known speech in Calcutta in 1786. So in this respect there was nothing original in Young's grouping. What was original to Young was the name he coined for the group: "Indo-European". This name (initially introduced in his review of Adelung in 1813) has stuck, and it soon lent itself to all manner of speculations upon how such exceptionally far-flung languages could originally have been closely related. As the archaeologist Colin Renfrew notes in his *Archaeology and Language: The Puzzle of Indo-European Origins* (1987), after nodding to Young's 1813 review: "what is the historical reality underlying this relationship? Where did these languages come from? Did they derive from a single group of people who migrated? Or is there an entirely different explanation? This is the Indo-European problem, and the enigma which has still not found a satisfactory answer."

The other two encyclopedia articles in Young's 'top three', apart from "Egypt", "Bridge" and "Tides", have fared somewhat less well with posterity than his "Languages". For example, the current *Encyclopaedia Britannica* gives substantial space to Young's modulus of elasticity in its treatment of engineering but does not mention his work on bridgebuilding, and contains no reference to his theory of tides; and this is true of other encyclopedias dealing with Young. The neglect was evident even in Young's own time, at least partly because of his insistence on anonymity in the case of "Bridge", which was written in 1816-17; "Tides", written in 1823, appeared under Young's name. When Sir George Biddell Airy, the astronomer royal, wrote on tides some years after Young's article, he completely overlooked Young, "although I well knew" (Airy later confessed to George Peacock) "that in writing on *any* physical subject it is but ordinary prudence to look at him first."

Peacock much admired Young's work on tides, devoting over ten pages to it in his biography—though more for its bold intuitions than for its mathematical elegance. When Airy at last took the trouble to read it, he too was impressed and admitted that Young's work deserved to have priority over his own in certain respects. Lord Rayleigh, speaking at the end of the nineteenth century, commented that, "In the theory of tides [Young] made great advances, and in explaining the circumstances which determine whether there will be high or low water under the moon, he gave the general theory of forced vibrations." Twentieth-century experts on hydrodynamics, such as Sir Horace Lamb, continued to salute Young's achievement in this area.

Tides are a difficult subject, even for specialists, and we shall not venture into it. In barest outline, Young abandoned Newton's oversimplification of the earth as a rotating solid sphere *entirely* covered with water, in which the tides were accounted for only by the gravitational forces acting between the ocean, the sun and the moon. Instead, Young categorized ocean waves into two types: 'forced' vibrations of water, induced by the gravitational forces of the heavenly bodies, and 'natural' vibrations, which followed spontaneously the laws of oscillation of water. He then treated the two vibrations as if they were two interacting pendulums, one of which had a period of oscillation dictated by the movement of the sun or the moon, and the other of which had a 'natural' period. So successful was Young's theory that he could make predictions about the tides in canals and narrow seas, which had been totally ignored in Newton's theory; as Airy noted, "he has hinted at the cause of the rapid rise of river tides as distinguished from their slower fall."

With "Bridge", Young was stimulated by the confusing submissions from various eminent men of science and engineers to a questionnaire compiled by a House of Commons committee examining a proposal to replace the old, 19-arched London Bridge (dating from 1176)—the sole bridge across the River Thames—with a modern cast-iron bridge designed by the celebrated bridge builder Thomas Telford, having a single arch spanning six hundred feet. In his article, Young tried to answer the committee's list of 21 questions as scientifically as possible, and in so doing he "deduced some important general principles on the statics of a masonry arch", notes Oldham, who also quotes Sir Charles Inglis, a twentieth-century expert on bridge construction at Cambridge University: "I was surprised to find how far [Young's] investigations had taken him and the accuracy of his conclusions. ... As a result of reading Young's article 'Bridge' I have realized that he had a mentality of the highest quality." But although Young's article came down essentially in favor of Telford's proposal—"the only reasonable doubt relates to the abutments"—the scheme was abandoned and a totally different, more orthodox design was adopted in the mid 1820s and completed in 1831.

The practical emphasis of many of Young's *Encyclopaedia Britannica* contributions chimed with his increasing involvement in public affairs during the same period and his diminishing hopes for his medical practice. In the last decade or so of his life, as we shall now see, he would become almost as much a man of the world, an adviser to institutions and the Government like the later Newton, as a man of science and a scholar. The new trend in his life began with an official request from the Admiralty for advice about how best to build wooden ships.

Chapter 13

In the Public Interest

"The cultivation of abstract science [is] obviously of far less impor-
tance than the preservation of the lives and property of seafaring
persons."

Young, report to the Admiralty as superintendent of the
Nautical Almanac, 1829

A round 1810, five years after the sea battle of Trafalgar had scup-
pered Napoleon's hopes of invading Britain, a master shipwright in
the Royal Navy dockyard at Chatham, Robert Seppings, proposed a
radical new method of building ships. His goal was to give a ship's structure
more strength and reduce 'arching'—its tendency to become convex
upwards, in the direction of its length—which sometimes broke a ship's
back or crippled it, especially during the process of launching. Seppings's
key innovation was to introduce diagonal trusses in the ship's construction.

Young's biographer Peacock explains the status quo and the new proposal well:

> The principal timbers and plankings of our ships were formerly dis-
> posed at right angles with each other. Thus the ribs were at right
> angles to the keel or backbone; the planks, both within and without,
> at right angles to the ribs; the beams which supported the decks at
> right angles to the outer framework ... By this arrangement of the
> timbers, the bolts which secured them to each other were generally
> found at the angles of a parallelogram, a figure which could collapse,
> or tend to collapse, without bringing into operation the strength of

its sides either to resist their compression or extension; but if diagonal beams are introduced and bolted into the sides at the opposite angles, the system becomes thenceforward firm and immoveable.

The reaction was extremely hostile—both from Seppings's fellow master shipwrights and from senior Royal Navy commanders. Most shipwrights and naval officials disliked innovation and the disruption of their perquisites. "I have been excommunicated by those in my own profession: indeed, they have passed judgment without making themselves masters of the principle. ... [It] accords with the old adage, 'Two of a trade can never agree'," Seppings complained to Young. Most commanders objected to the fact that Seppings wanted to substitute round sterns for the traditional, if vulnerable, flat ones, because round sterns were both stronger and more easily armed with cannon. "Many of the old captains and admirals, whose magnificent stern drawing-rooms were [to be] thus invaded by 32-pounders, were furious in their opposition," notes Peacock.

But the lords of the Admiralty, and especially one of its secretaries John Barrow, were impressed by Seppings. They had him brace a 74-gun ship, H.M.S. *Tremendous*, and put it on trial for many months in the North Sea. Seeking scientific support for the change, in 1811 Barrow invited Young, among other men of science, to report on the Seppings proposals to the Admiralty. For Young had included carpentry in his Royal Institution lectures and had, of course, discussed mechanical strength in detail in his *Natural Philosophy*.

Young was hesitant to accept, given his renewed commitment to medicine; but the national importance of the issue seems to have swayed him, and no doubt he was somewhat flattered to be asked for his considered opinion by Government. He replied to Barrow:

I ought perhaps to have returned an earlier answer to your official letter, but I have made so many resolutions to forswear all further concern with the mathematical sciences, that I could not at once determine again to deviate from them by accepting their lordships' invitation. Recollecting, however, that as far as I know, I am the only person in this country that has communicated to the public any attempts to improve the *theory* of carpentry (*Lectures*, Chap. XIV.), and that it would be scarcely decent to draw back on an occasion where I was called on to assist in a case of *practical* importance, I have overruled my hesitations ...

After viewing Seppings's designs, Young reported to the Admiralty hardly more than a month later, at the end of 1811, and in 1814 he published the substance of his report in the *Philosophical Transactions* of the Royal Society under the title, "Remarks on the employment of oblique riders and on other alterations in the construction of ships". His analysis involved serious science and considerable mathematics, and its support for Seppings was distinctly qualified. In section nine, "Mr Seppings's braces", Young notes: "It appears, therefore, to be sufficiently established, that the principle of employing oblique timbers is a good one, provided that it be so applied as to produce no practical inconvenience. We must next inquire whether Mr Seppings has introduced it in a manner likely to be effectual, and not liable to any material objections." In the final section, under "Conclusion", Young states:

> It is by no means impossible, that experience may suggest some better substantiated objections to these innovations, than have hitherto occurred: but none of those objections which have yet been advanced, appear to be sufficiently valid to warrant a discontinuance of the cautious and experimental introduction of Mr Seppings's arrangements, which has been commenced by orders of the board of Admiralty.

Although Young, without doubt, was basically in favor of the new method of construction, his report was hardly a ringing endorsement of it. "Dr Young was not easily seduced into enthusiasm," observes Peacock. An embarrassed Admiralty official who vehemently opposed the changes informed Young: "Though science is much respected by their lordships, and your paper is much esteemed by them, it is too learned." Barrow, as the main booster of Seppings, felt marooned. No scientist himself but an able popularizer of science (and later a founder of the Royal Geographical Society), Barrow was critical of Young's report in the *Quarterly Review*:

> He cannot, we think, disapprove of the principle; yet so many conditionals, hypotheticals, and potentials are employed, that if approbation be meant, either of the principle or its application, it is at any rate 'damn'd with faint praise'. Dr Young will not infer from this that we undervalue science ... our regret arises from seeing 'abstract science' misapplied, in raising doubts on points of practice which common sense and experience are best able to determine, and which no calculus can reach.

At any rate, Seppings and Barrow got their way. The new methods of shipbuilding were introduced and became established; Seppings was appointed surveyor of the navy in 1813 and elected a fellow of the Royal Society; and in due course both he and Barrow received knighthoods.

In 1814, Young became a member of a committee of the Royal Society appointed at the request of the Home Department of the Government to examine the introduction of coal gas into London. Two years previously, the Gas-Light and Coke Company had been chartered to light the City, Westminster and Southwark, and soon rival companies followed with gas-lighting contracts north and south of the river (including Young's home area in Welbeck Street). For reasons of cost, the gasworks were located mainly by the river, but they were nonetheless close to areas of dense population. One of them, at Woolwich, had recently exploded, causing general alarm and the setting up of the Royal Society committee. Its purpose was to establish how dangerous coal gas really was: through what length of tubing, if any, would its flame run back and ignite the reservoir?

Young was not involved in the chemical part of this inquiry, which clearly showed that the flame of gas in a small tube is not transmissible. (In the following year, 1815, this led to Sir Humphry Davy's development of the miner's safety lamp for preventing methane explosions in mines, after Davy returned from a long tour of Europe with Faraday.) Young's role was to investigate the likely force of a gas explosion by comparing it with the explosive force of gunpowder. "It was a problem of a very high order of difficulty, the elements of whose solution were of a very hypothetical and uncertain character," writes Peacock. Young concluded that the explosive force of coal gas, when mixed with atmospheric air, was "somewhat less than one-thousandth part of the same mass of gunpowder." Therefore, with due precautions, it was safe to site gasworks within London—however ugly the gasometers might be.

Young had agreed to write on "Weights and measures" for the *Encyclopaedia Britannica*, and it was only natural that he should become involved with official inquiries into the need for increasing standardization as science progressed. In 1816, he was appointed secretary to a government commission for ascertaining the length of the seconds pendulum; for comparing the French decimal and metric system (adopted in the 1790s during the French Revolution) with English units; and for considering the practicability and advisability of converting to a more uniform system of weights and measures throughout the British Empire (what would in due course become known as 'imperial' units). The commission—whose members included the president of the Royal Society Sir Joseph Banks, the physicist and chemist William Hyde Wollaston and Davies Gilbert, a future president of the Royal Society—produced three reports, in 1819, 1820 and 1821, and these were drawn up by Young; they formed the basis of his encyclopedia article.

With the exception of the seconds pendulum, the commissioners avoided making any recommendations that smacked of revolution. They suggested only the introduction of an 'imperial gallon', and were opposed to decimalization of the coinage and the primary units of weights and measures, preferring to retain the prevalent duodecimal system and other scales (for example, pounds, shillings and pence; miles, yards, feet and inches). Peacock, writing in the 1850s, was thoroughly disapproving of this decision; he noted that an 1837 commission had taken the opposite view, and declared that decimalization would soon be legislated for! In reality, it would take until 1971 for the British currency to go decimal, and longer still for British weights and measures to go metric.

While Young was certainly in favor of precision in science, he was conservative in recommending changes that would deeply affect the public. Hudson Gurney, who was elected an M.P. in 1816, commented in his posthumous memoir of his old friend:

> It seems right to state, that in pursuing these investigations it was his opinion, that however theoretically desirable it might be, that all weights and measures should be reducible to a common standard of scientific accuracy, yet that, practically, the least possible disturbance of that to which people had long been habituated was the point to be looked to, and on this ground he was extremely averse to unnecessary changes.

With the definition of the length of the seconds pendulum, Young the scientist came to the fore. The idea had been around for many decades that the time period of a swinging pendulum could be used to define a standard yard (or meter). The reason is that the period of a simple pendulum depends not on the mass of its bob, but only on the length of the pendulum; thus all pendulums of the same length swing at the same rate. "The length of a simple pendulum which beats seconds is therefore a 'natural' standard of length which can be constructed anywhere," writes Frank Oldham—from which the length of a meter can be calculated. (In practice, adjustments have to be made for temperature, which affects the length of the pendulum, and any local variations in the earth's gravitational field due to the presence of large masses such as mountains.)

In the 1740s, writes Young in his encyclopedia article, "George Graham, the watchmaker, determined ... the correct length of the pendulum vibrating seconds to be 39.130 inches". By 1814, in the report of a Parliamentary committee, this length had been refined to 39.13047 inches and the length of the meter to 39.3828 inches, which Young, after adjustment for different standard temperatures in Britain and France, further refined to 39.3710 inches. Then, during the time of the commission appointed in 1816, Captain Henry Kater, another of the commission's members, devised and constructed a new pendulum—"with great ingenuity ... and great mechanical skill" wrote Young—consisting of a bar pivoted on two knife edges. The bar was pivoted from each edge in turn and the positions of movable weights were adjusted so that the period of the pendulum was one second with both pivots. The distance between the knife edges was then known to be the length of the seconds pendulum: obviously a more accurate measurement than the measurement obtainable from a suspended bob pendulum. In 1818, Kater published a paper on his pendulum, to which Young appended a note. The resultant figure for the seconds pendulum was now even more accurate: 39.13929 inches, with the meter given as 39.37079 inches, and these figures were included in the final report of the commission, laid before Parliament in 1824.

Around this time, late in 1818, Young received some surprising news. One morning he opened a newspaper and discovered that his name had been proposed, without his knowledge, in the House of Commons to be one of three 'resident' commissioners (i.e., resident in London) of the newly reconstituted Board of Longitude, the other two being Wollaston and Kater. Davies Gilbert, who was an M.P., was probably responsible for the nomination. In the event, instead of becoming a commissioner, Young was cajoled into becoming secretary of the board at an annual salary of £100 and superintendent of the board's *Nautical Almanac* at an additional salary of £300, though he would have preferred to take only the post of superintendent. "There could be no doubt that such an engagement, though it occupied little of his time, was likely to interfere very much with his medical reputation," Young wrote of himself in his autobiographical sketch, "but as he never had a family, he was not ambitious to acquire a large fortune, and preferred a competence well secured, to a contingency of greater affluence: and from this time forwards, he considered his salary, together with his own and his wife's property, as affording such a competence." In other words, when he took on the two appointments at the Board of Longitude in 1819, he rang the death-knell for his medical ambitions—though it would take him another four years before he agreed to put his name to all his publications.

But if the two posts brought Young a degree of financial security in middle age, they brought him no peace of mind. Instead, both positions embroiled him in stormy controversy—as stormy as that with the *Edinburgh Review* in 1804 and much longer lasting. As in the 1760s and after, when the members of the original Board of Longitude (set up in 1714 in Newton's day) refused to pay the watchmaker John Harrison a promised award for his invention of an accurate marine chronometer (as described in Dava Sobel's bestselling *Longitude*), so in the 1820s, a group of hostile astronomers ganged up on Young as secretary and eventually succeeded in having the Board of Longitude abolished in 1828. Until his dying days, Young would be locked in very public, and often acrimonious, dispute with these astronomers.

At the heart of the argument was the *Nautical Almanac*, started in 1767 by the astronomer royal—Harrison's nemesis—Nevil Maskelyne, who maintained it for more than four decades until his death in 1811. It was a compendium of astronomical tables and navigational aids, which included many of the results of Maskelyne's studies of the heavenly bodies: the sun, the moon, the planets and the stars. And of course it needed regular updating with astronomical data computed to a high accuracy for several years into the future. Both the seafaring captains and the earth-bound astronomers of an island nation found the almanac invaluable in their observations and calculations. But the data and presentation that each group wanted were not always the same (moreover, the almanac came to include serious errors after Maskelyne's death). The question increasingly arose: who was the *Nautical Almanac* primarily intended for—seamen or astronomers? Young and the Admiralty answered: seamen. The astronomers, who were then energetically forming their own professional association, separate from the Royal Society, the Astronomical Society of London—which shortly became the Royal Astronomical Society—almost inevitably prioritized themselves over seamen.

Peacock, who was Lowndean professor of astronomy at Cambridge, was inclined to be critical of Young as superintendent of the *Nautical Almanac*. "It is not easy," he wrote, "to define the precise limits which separate the wants of the navigator from those of the traveler and astronomer. A scientific and well-educated captain may be placed under circumstances which will require him to act in all these capacities, when he visits unknown regions". Peacock thought Young generally too rigid in his distinction between the needs of navigators and of astronomers, and personally somewhat unsympathetic to astronomers, especially those behind the new astronomical society. According to Peacock's final verdict:

> [I]t is difficult for the warmest admirers of Dr Young altogether to justify the line of conduct which he pursued. Of the two grounds upon which he chiefly rested his defense—expense to the Government, and the interests of navigation—the first was absolutely unworthy of notice, and the second could hardly be compromised by the embarrassment produced by placing in the hands of seamen more than they required, when the most simple instructions would direct them what to look for.

Alex Wood's biography, at a century's remove from Peacock's, is more inclined to give Young the benefit of the doubt. A substantial appendix devoted to the controversy, written by Edmund Dews, an Oxford academic, notes that American seamen preferred to use an abridged edition of the British almanac until the first publication of an American almanac in 1852, which came in two editions, one for seamen and the other for astronomers. The papers of the Board of Longitude, whose members included distinguished astronomers, show diligent supervision of and consistent support for Young's superintendence. In the campaign for reform of the board, "unscrupulous methods" were used by certain astronomers, writes Dews. He concludes: "The price finally paid for the 'reform' of the almanac was the abolition of the Board of Longitude and the stopping for twenty years of regular provision for Government support of research in the physical sciences."

The debate still rumbles. A current historian of science at Cambridge University, Simon Schaffer, contributing to a history of the Lucasian professors of mathematics at Cambridge, over-confidently assumes that Young was some kind of dinosaur, unable to move with the scientific times. (One is reminded of the underestimation of Young by the over-confident Emmanuel College tutor who knew him as a colleague in the 1790s.) Schaffer writes that George Biddell Airy, the Lucasian professor, who in 1828 was also appointed professor of astronomy and director of the Cambridge Observatory (and later the astronomer royal), became a member of the Board of Longitude in 1827-28 and acted as a new broom. Airy "headed a reform campaign against the regime of the conservative natural philosopher Thomas Young. ... By spring 1828 the board had been abolished for expensive inefficiency, and Airy lobbied hard to take over its almanac." But this unsympathetic picture of Young does not square easily with Airy's support for Young in his own autobiography and with Young's irritated but nevertheless reasonable letter to Airy written at the time, around 1828:

> If every practical astronomer were like you, I should think it right for the Admiralty to consider the importance of saving your time almost as much as that of nautical men. But when I see people who possess nothing of science but a few fine instruments and a good deal of leisure, affecting to call themselves astronomers and to dictate to the public what ought to be done for the promotion of astronomical science, I do certainly feel a disposition to rebel against their authority ... With respect to the N. A., I hope I so expressed myself as professing a readiness to be convinced by you and not to adopt your opinions without having vanquished my own doubts. I am most anxious for your assistance in recommending whatever you think right, and I trust you will not condemn me, if I am not always persuaded.

As for the abolition of the Board of Longitude, even the skeptical Peacock was outraged: he called it "an act of barbarism which was neither called for by any just considerations of expediency nor of rational economy." While Young's French physicist and astronomer friend Dominique

Arago—incidentally a man of the extreme left in French politics, about as far from being a conservative as it was possible to be—condemned the British Government's decision in the most impassioned terms in a eulogy given before the French National Institute after Young's death:

> An orator … who had hitherto vented his spleen only on productions of French origin, attacked the most eminent men of England, and uttered against them, before Parliament, the most puerile accusations with laughable gravity. The ministry, whose eloquence was exercised for whole hours upon the privileges of rotten boroughs, did not utter a single word in favor of genius; and, finally, the Board of Longitude was suppressed without opposition. … The learned secretary, at least, should not have been separated from his colleagues; nor should this sensitive individual, rich in all the fruits of human intelligence, have been rated before the representatives of his country, like so much sugar, coffee, or pepper, in pounds, shillings, and pence.

We shall return to the outcome of the battle in the last chapter of this book. For now, it would be wrong to leave the impression that Young spent his entire period as secretary and superintendent engaged in this struggle. Some of his work was considerably more rewarding and constructive. Here are just two examples.

As secretary of the Board of Longitude, in 1820, Young successfully pushed for the government to establish a permanent observatory at the Cape of Good Hope in South Africa, on the grounds that it would be "highly conducive to the improvement of practical astronomy and navigation". This soon became an important observatory, in collaboration with the separate observatory near Cape Town established by Sir John Herschel in the 1830s.

Another duty was to verify claims by sailors to have discovered the North-West Passage through the Arctic linking the Atlantic and the Pacific Ocean, for which the Board of Longitude offered a range of awards from £5-15,000, depending on how far north and west a ship managed to reach. In November 1820, an award of £5000 was made by the board to Lieutenants William Edward Parry and Matthew Loudon for sailing within the Arctic Circle in the summer of 1819, even though they had not found a complete passage. Young's assessment of this expedition prompted an interesting letter from him to Gurney:

And here is the *polar expedition* arrived, whom I am to examine on their oaths to get them the £5000, which it seems will be spent on lowering the price of oil, by the information they have given the whalers. I imagine also they have set the practical question of the passage at rest, as it is obvious that there would be no reasonable chance of getting to Behrings' Straits in the short Arctic summer of six weeks, even if there is a passage, which seems by no means improbable ... I should not, however, be surprised if the curiosity of the Admiralty prompted them to continue the research, and I have no objection to curiosity in others, though it is a great many years since I was scolded for that quality myself.

In making this self-criticism, tongue in cheek, Young was probably thinking of his youthful exploits on a horse through the Scottish Highlands and of his travels on foot and by coach through Germany as a student, cut short by the war with Napoleon. For a man of such global interests, who was foreign secretary of the Royal Society, Young had actually traveled comparatively little. Now, before it was too late, he decided to make good on his youthful promise to himself, and undertake a grand tour in Europe.

Chapter 14

Grand Tour

"[Our expedition] seems like the last act of my boyhood and the first of my old age: on the one hand a sort of finish to my Latin and Greek, and on the other, a setting at defiance all professional conveniences in a way which may be deemed somewhat imprudent in a servant of the public. But I do not owe the public much, and I suppose I shall never be paid much of what the public owes me."

Young, letter to Hudson Gurney, 1821

Since leaving Germany as a medical student in early 1797, Young had paid one visit to France during the brief Peace of Amiens in 1802—when he had heard Napoleon speaking at the National Institute in Paris—and two visits to Paris in 1817—when he was welcomed at the same institution by Arago and other eminent scientists, following Arago's visit to Young in England in 1816 (with Gay-Lussac) to discuss the undulatory theory and Fresnel's experiments. This was the sum total of Young's travel in Europe, or indeed anywhere else outside Britain, during those two decades of war between Britain and France.

Now, having established himself at the Board of Longitude and more or less given up medicine, except for his rounds as a physician at St George's Hospital, "in the summer of 1821"—to quote Young's autobiographical sketch—"meaning to discontinue his professional residence at Worthing, he took the opportunity of making a hasty tour of Italy, which he considered as a part of his education that had before been unavoidably postponed. In about five months, he saw all the remarkable cities of Italy, and returned by Switzerland and the Rhine." His wife Eliza accompanied him throughout; the two of them were away from London from the mid- dle of June until the end of October.

One would hardly expect of Young, who was also representing the Royal Society as its foreign secretary, that his European tour would be a conventional sightseeing holiday. And one would not be wrong. Along with visiting the monuments and artworks of classical and Renaissance Italy in Turin, Genoa, Rome, Naples, Sienna, Pisa, Florence, Venice and Milan, and meeting friends and acquaintances who were living abroad as well as distinguished strangers, Young blended some serious science with ancient Egyptian epigraphy and study of Italian painting and sculpture. In Rome, for instance, guided by a friend he had known in his Cambridge days, the traveler and archaeologist Edward Dodwell, who was resident in the city, Young was keen to see the Egyptian obelisks erected by the Romans, which had so exercised the imagination of Athanasius Kircher in the seventeenth century. He was interested in the different modern attempts at restoration, some awkward, others more successful, like the Lateran obelisk that included a "block of granite, which … still exhibits some words of a Latin inscription, turned upside down, but not effaced, although the hieroglyphics belonging to the place have been imitated with tolerable fidelity."

The first stop for the Youngs was naturally Paris, where he arrived in time to attend a meeting of the National Institute, which was then, admits his biographer Peacock, "by far the most illustrious scientific body in Europe". Whether Britain and its Royal Society were part of Europe or not according to Peacock, he does not clarify, but either way his statement is true of this period in science. In Newton's time, British science had led the world; the early nineteenth century was unquestionably the era of French scientific dominance.

Young was greeted cordially by Arago and Alexander von Humboldt, the great naturalist and explorer, and introduced to Laplace, the palaeon- tologist Georges Cuvier, the astronomer and physicist Jean-Baptiste Biot and other major scientists—but not to Jean-François Champollion, who was still living in Grenoble in obscurity, poverty and poor health, as a result of his support for Napoleon in 1815. (In July, by sheer chance just after the departure of Young from Paris, and two months after the death of

Napoleon in St Helena, Champollion finally took the plunge and returned to the capital, where a year or so later he and Young would at last meet at a pivotal moment in the decipherment, as we shall see in Chapter 15, "Dueling with Champollion".) In mid 1821, Young's optical work was known to the National Institute through the strong advocacy of Arago, but Fresnel had yet to publish his full theory of light as a transverse wave. Therefore, although Young was undoubtedly treated as an honored guest in Paris, neither of his pioneering contributions—to the wave theory of light nor, to a lesser extent, to the decipherment of the hieroglyphs—were by any means *faits accomplis*. It would take another six years, until 1827, before Young would be elected as one of the eight foreign associates of the National Institute (in place of the deceased physicist Alessandro Volta)—the only scientific honor that he seems really to have coveted.

The Youngs now headed for Italy, traveling by coach via Lyon and Chambéry through Savoy and over the Alps through the Mount Cenis Pass to Turin—following, by an odd twist of history, almost exactly the same route at almost exactly the same time as Champollion journeying in the opposite direction from Grenoble to Paris. They were excited by their first encounter with Alpine scenery. From Novi, beyond Turin on the way to Genoa, Young told Gurney on 8 July: "We were delighted beyond measure with Savoy, a country which seems too little known to travelers in comparison with Switzerland, at least it far exceeds any idea which I had formed of Switzerland without having seen it. Turin is a most magnificent city ..."

Two years later, Turin would have been even more attractive to Young, for its museum had by then acquired some Egyptian treasures purchased by the king of Sardinia-Piedmont, which Champollion would hasten to study in 1824. They came from the vast collection of Bernardino Drovetti, the former French consul-general in Egypt. In the autumn of 1827, the rest of Drovetti's collection would be purchased by the king of France at the instigation of Champollion (who exulted over "jewelry of an unbelievable magnificence" and objects carrying royal inscriptions, including "a cup in solid gold"), and added to the Louvre Museum in Paris. During this decade or two, collectors of Egyptian art had a field day in selling their collections. Egypt was all the rage, fueled by such displays of Egyptian art as found in the settings of Mozart's opera *The Magic Flute*. The collections of the English consul-general in Egypt, Henry Salt, of the Swedish-Norwegian consul-general, Giovanni Anastasi, and of Drovetti's archrival, Giovanni Belzoni, along with those of others, together laid the foundations of today's great Egyptian galleries in London, Paris, Bologna, Florence, Turin and Leiden. The Egyptologist Sir Alan Gardiner commented in 1961: "the excavations exploited or instigated by [these early collectors] were little better than lootings, though their authors should not be condemned for disregard of scientific standards not yet born."

Even so, in 1821, Young was determined to view Drovetti's collection, which had for a long while been warehoused at Leghorn (Livorno) on the western coast of Italy near Pisa, after being transported from Egypt. Having gained access, he made a dramatic discovery—another bilingual inscription, so far largely unknown to scholars. From Florence, he wrote to Gurney on 8 September:

> Pisa amply repaid us for taking this circuitous route; Leghorn, if possible, still more. But what you will be pleased to hear, is the discovery that I made of a bilingual stone among Drovetti's things, which promises to be an invaluable supplement to the Rosetta inscription as I dare say Drovetti is well aware. There are very few distinct hieroglyphic characters about the tablet, and the rings [cartouches] for the names of the king are left blank: but there are one or two well-known personages of the Egyptian pantheon whom I shall be glad to find named in Greek, and the blank names can be of little consequence as they must have been some of the dynasty of the Ptolemies, and I think there are some emblems of Ptolemy Philopator. Under the tablet are about 15 lines of enchorial [demotic] character, and about 32 in Greek, not at all distinctly legible, but nowhere totally effaced, so that I believe that with care every part of the inscription may be recovered.

The problem, of course—a problem that has to be faced by all would-be decipherers of other people's inscriptions—was how to obtain an accurate copy. Young now shared his plan with Gurney:

> I could not get leave to take a copy, the merchant having no authority to do anything beyond the safe custody of the collection. But he has consented that I should send an experienced artist from Florence to take two casts, or rather impressions of the stone, one or both of which I hope Drovetti will let me have for myself or for the [British]

Museum on fair terms: but if he does not, I have only stipulated that whenever the collection is embarked, the copies shall remain safe at Leghorn until it has arrived at the place of its destination without injury from shipwreck or other accidents; and I shall have the satisfaction of thinking that I have at least done something for the preservation of the second great treasure of Egyptian literature, which is so far of infinitely more consequence than the first, as I suppose there are no good copies of it yet in existence: and if the original were lost without a copy, we should lose the means of confirming or correcting and perhaps greatly extending what the Rosetta Stone has already enabled us to establish.

In his letter to the Pisan merchant setting out this proposal, Young had written: "Whatever may be Mr Drovetti's decision, I trust that this application, from one who flatters himself that he is the only person living, that can fully appreciate the value of the object in question, will at least not be disagreeable to him." Perhaps this was a shade presumptuous, but there is no doubt that Young was overwhelmingly the leader in this field in 1821, when Champollion was barely known. Anyway, the interests of scholarship did not, alas, prevail over those of commerce: the merchant was willing to help, but Drovetti was not. As Young noted in early 1823, with justifiable asperity, in his book, *An Account of Some Recent Discoveries in Hieroglyphical Literature and Egyptian Antiquities*:

Mr Drovetti's cupidity seems to have been roused by the discovery of an unknown treasure, and he has given me to understand, that nothing should induce him to separate it from the remainder of his extensive and truly valuable collection, of which he thinks it so well calculated to enhance the price; and he refuses to allow any kind of copy of it to be taken.

But, as it often happens to those who are too eager to monopolize, he has now outstood his market, and the pearl of great price, which six months ago I would have purchased for much more than its value, is now become scarcely worth my acceptance. I was principally anxious to obtain from it a collateral confirmation of my interpretation of the enchorial inscription of Rosetta; but having fortunately acquired materials, from other sources, which are amply sufficient for this purpose, I can wait, with great patience, for any little extension, which my enchorial vocabulary might receive from this source.

The conclusion of this particular tale must await the last chapter, which deals with Young's final researches on the enchorial/demotic script.

Most of the Italian tour had nothing to do with Egypt, of course. Though Young said nothing of Italian food or wine (or, for that matter, Italian women), he clearly loved being in Italy, like so many Englishmen before and since. "On the whole our expedition has been extremely prosperous," he wrote to Gurney from Florence in early September, "and like most other things I have done in life I am very glad that I have done it, though I am by no means certain that I should have resolution to do it again." Naples "delighted us extremely; and our expedition to Paestum repaid us better than I had expected; but much more by the beauty of the scenery about Salerno than by the magnificent copies of the cork models which we had seen in London, for the ruins seem to be perfect imitations of pieces of cork on a large scale." In Rome, "We fell into a good deal of society ... chiefly among the diplomatic people." (His friend Dodwell was close to the pope.)

As for Italian intellectual and artistic life, he was sympathetic but critical:

> Of the science and literature of this country I know nothing: but I cannot help fancying that Pozzo di Borgo must have been dreaming when he told me at Paris that the Italians were making great strides in the improvement of the human intellect. I do not think there are any living poets of transcendent merit: none certainly to rival some of ours: in painting they have nothing but a few good draughtsmen and copyists: in sculpture, they have Canova who probably comes next to Michelangelo; they have Thorwaldsen, Bartelini, and a few others, about as good as our own: their taste for music seems to be altogether exhausted, and we sought in vain for a little harmony at St Peter's and in the pope's choristers. At Naples, however, their opera and their ballet is well mounted: and the theater of San Carlo illuminated was the most magnificent spectacle I ever beheld.

Unfortunately for the Youngs, there was bad news waiting for them at Schneiderff's Hotel in Florence. Mrs Maxwell, Eliza's mother, was seriously ill in England. Young told Gurney that they must now "hasten home to take our share in the attentions which she is entitled to receive from her family; and unless we get rather more favorable letters at Milan we shall certainly be in London this day six weeks [hence], provided no unforeseen accident should detain us." Their visit to Venice was therefore much briefer than they would have liked. Their plan was to stick to their original route as far as Geneva, "and then to give up Switzerland and the Rhine if it should be necessary, and to return by Dijon, Troyes and Lille." But when they got to Geneva, the information reached them that Mrs Maxwell had died. The grand tour was effectively over. The travelers proceeded rapidly through Switzerland to Schaffhausen, and then through the valley of the Rhine. On 21 October, they were in Brussels, from where Young wrote a final letter to Gurney, before reaching London at the end of the month.

From Florence, Young had written to his old friend that the tour seemed to mark a natural break in his life, between "boyhood" and "old age". No doubt the subsequent death of his wife's mother, who was probably not much more than ten, or at most fifteen years, older than him, had the tendency to confirm his self-diagnosis. Though he would visit the continent again, more than once, he would travel only as far as Paris, Belgium, Holland and Geneva, and only for relatively brief periods. The rest of his life would be spent entirely in London, at the vortex of the capital's scientific, literary and political life.

Chapter 15

Dueling with Champollion

"Mr Champollion, junior … has lately been making some steps in Egyptian literature, which really appear to be gigantic. It may be said that he found the key in England which has opened the gate for him, and it is often observed that c'est le premier pas qui coûte [it is the first step which costs]; but if he did borrow an English key, the lock was so dreadfully rusty, that no common arm would have had strength enough to turn it."

Young, letter to William Hamilton, 1822

When Young returned to London from his grand tour in late 1821, a highly dramatic new phase in ancient Egyptian researches was about to begin. In the first phase, from the discovery of the Rosetta Stone in 1799 until the publication of his *Encyclopaedia Britannica* article, "Egypt", in 1819, Young had had the field of hieroglyphic decipherment largely to himself. Now he would be joined in earnest by Champollion, who would quickly overtake him and become the founder of Egyptology as a science.

During the 1820s, the two men sometimes cooperated with each other, but mostly they competed as rivals. Their relationship could never have been a harmonious one. Young claimed that Champollion had built his system of reading hieroglyphics on Young's own discoveries and his hieroglyphic 'alphabet', published in 1815-19. While paying generous and frequent tribute to Champollion's unrivaled progress since then, Young wanted his early steps recognized. This Champollion was adamantly unwilling to concede, and in his vehemence he determined to give all of

Young's work the minimum possible public recognition. Just weeks before
Young's death in 1829, Champollion, writing in the midst of his expedition
to ancient Egypt—he was then at Thebes in the Valley of the Kings (a place
he had just named)—exulted privately to his brother:

> So poor Dr Young is incorrigible? Why flog a mummified horse?
> Thank M. Arago for the arrows he shot so valiantly in honor of the
> Franco-Pharaonic alphabet. The Brit can do whatever he wants—it
> will remain ours: and all of old England will learn from young France
> how to spell hieroglyphs using an entirely different method ... May
> the doctor continue to agitate about the alphabet while I, having been
> for six months among the monuments of Egypt, I am startled by
> what I am reading fluently rather than what my imagination is able to
> come up with.

The nationalistic overtones—at times evident in Young's writings,
too—have to some extent bedeviled honest discussion of Young and
Champollion ever since those Napoleonic days of intense Franco-British
political rivalry. Even Young's loyal friend, the physicist Arago, turned
against his work on the hieroglyphics, at least partly because Champollion
was an honored fellow countryman. Thus, a recent French book for the
general reader by a writer of Egyptian origin, Robert Solé, and the
Egyptologist Dominique Valbelle, translated into English as *The Rosetta
Stone: The Story of the Decoding of Hieroglyphics*, deliberately omits the
trenchant criticism of Champollion's character written to Young in 1815
by his former teacher Sylvestre de Sacy (who would hail Champollion for
his success ten years later), quoted earlier; it also omits two other controversial episodes, in which Champollion is generally held to have suppressed an erroneous publication of his own and to have failed to
acknowledge a crucial inscriptional clue provided by another. (We shall
come to these in more detail.)

Alongside this, Egyptologists, who are the people best placed to understand the intellectual 'nitty-gritty' of the dispute, are naturally drawn to
Champollion more than Young, because he founded their subject. No
scholar of ancient Egypt would wish to think ill of such a pioneer. Even
John Ray, the Egyptologist who has done most in recent years to give
Young his proper due, admits: "the suspicion may easily arise, and often
has done, that any eulogy of Thomas Young must be intended as a

denigration of Champollion. This would be shameful coming from an Egyptologist."

Then there is the cult of genius to consider: the fact that many of us prefer to believe in the primacy of unaccountable moments of inspiration over the less glamorous virtues of step-by-step, rational teamwork. Champollion maintained that his breakthroughs came almost exclusively out of his own mind, arising from his indubitably passionate devotion to ancient Egypt. He pictured himself for the public as a 'lone genius' who solved the riddle of ancient Egypt's writing single-handedly. The fact that Young was known primarily for his work in fields other than Egyptian studies, and that he published on Egypt anonymously in his first phase, made Champollion's solitary self-image easily believable for most people. It is a disturbing thought, especially for a specialist, that a non-specialist might enter an academic field, transform it, and then move onwards to work in an utterly different field.

Lastly, in trying to assess Young and Champollion, there is no avoiding the fact that they were highly contrasting personalities and that this contrast sometimes influenced their research on the hieroglyphs. Champollion had tunnel vision ("fortunately for our subject", says Ray); was prone to fits of euphoria and despair; and had personally led an uprising against the French king in Grenoble, for which he was put on trial. Young, apart from his polymathy and a total lack of engagement with party politics, was a man who "could not bear, in the most common conversation, the slightest degree of exaggeration, or even of coloring" (says Gurney). They were poles apart intellectually, emotionally and politically.

Consider their respective attitudes to ancient Egypt. Young never went to Egypt, and never wanted to go. In founding an Egyptian Society in London in 1817, to publish as many ancient inscriptions and manuscripts as possible, so as to aid the decipherment, Young remarked that funds were needed "for employing some poor Italian or Maltese to scramble over Egypt in search of more." Champollion, by contrast, had long dreamt of visiting Egypt and doing exactly what Young had depreciated, ever since he saw the hieroglyphs as a boy; and when he finally got there, he was able to pass for a native, given his swarthy complexion and his excellent command of Arabic. In his wonderfully readable and ebulliently human *Egyptian Diaries*, Champollion describes entering the temple of Ramses the Great at Abu Simbel, which was blocked by millennia of sand:

I almost entirely undressed, wearing only my Arab shirt and long underwear, and pressed myself on my stomach through the small aperture of a doorway which, unearthed, would have been at least 25 feet high. It felt as if I was climbing through the heart of a furnace and, gliding completely into the temple, I entered an atmosphere rising to 52 degrees: holding a candle in our hand, Rosellini, Ricci, I and one of our Arabs went through this astonishing cave.

Such a perilous adventure would probably not have appealed to Young, even in his careless youth as an accomplished horseman roughing it in the Scottish Highlands. His motive for 'cracking' the Egyptian scripts was fundamentally philological and scientific, not aesthetic and cultural (unlike his attitude to the classical literature of Greece and Rome). Many Egyptologists, and humanities scholars in general, tend not to sympathize with this motive. They also know little about Young's scientific work and his renown as someone who initiated many new areas of scientific enquiry and left others to develop them. As a result, some of them seriously misjudge Young. Not knowing of his fairness in recognizing other scientists' contributions and his fanatical truthfulness in his own scientific work, they jump to the obvious conclusion that Young's attitude to Champollion was chiefly envious. The classicist Maurice Pope says this more or less in his book, *The Story of Decipherment*, as quoted in the introduction; while two archaeologists, Lesley and Roy Adkins, in *The Keys of Egypt: The Race to Read the Hieroglyphs*, state openly that "while maintaining civil relations with his rival, Young's jealousy had not ceased to fester." Not only would such an emotion have been out of character for Young, it would not have made much sense, given his major scientific achievements and the fact that these were increasingly recognized from 1816 onwards—starting with French scientists. For Champollion, the success of his decipherment was a matter of make or break as a scholar; for Young, his Egyptian research was essentially yet another fascinating avenue of knowledge to explore for his own amusement.

Champollion's first significant publication on the Egyptian scripts came in April 1821, and appeared from Grenoble, just three months before he left that city for Paris. Nowhere in it did he make any reference to Young, and, according to Champollion, he was unaware of Young's published article "Egypt" until he returned to Paris. His *De l'Écriture*

Hiératique des Anciens Égyptiens consisted of a mere seven pages of text and seven plates. It announced four firm conclusions, of which two were important. One was correct: that the hieratic script on Egyptian manuscripts—and hence presumably the demotic script—was only a "simple modification" of the hieroglyphic. (Young had come to a similar conclusion in 1815 and published it in his "Egypt" in 1819.) The other conclusion was incorrect: that the hieratic/demotic characters were "signs of things and not of sounds"—in other words, there was *no* phonetic element in the hieratic/demotic script, which was a conceptual script like the hieroglyphs, said Champollion. (Young, and before him Åkerblad, of course was certain that there *was* an alphabetical element in the demotic, but that this element was mixed with non-phonetic signs derived from the hieroglyphs.)

The error was a serious one, and it seems as if Champollion soon realized this, because he is alleged to have made strenuous efforts to withdraw all copies of the 1821 publication, suppress the text and redistribute only the plates. The allegation is likely to have been true, given the subsequent rarity of the publication, the fact that Champollion presented only the plates to Young, who was unaware of the text, and, most telling of all, that Champollion chose to make no reference of any kind to the publication in his breakthrough publication of 1822. Clearly, in the year that elapsed between August 1821 (when Champollion lectured to the Academy of Inscriptions in Paris on the ideas in his Grenoble publication) and his announcement of a decipherment to the same institute in September 1822, Champollion changed his mind and decided that there was, after all, an alphabetic element in the Egyptian scripts. The question then becomes, what caused his change of mind?

It was now, in Paris in 1821-22, that Champollion definitely studied Young's article in the *Encyclopaedia Britannica*, by his own admission. Though the idea does not seem credible, Champollion asked the world to believe that the article did not substantially influence his thinking. But it would, without doubt, have made him aware of Young's belief in a phonetic element in the scripts, published in the form of a short list of hieroglyphs representing "Sounds" and a second list of demotic signs labeled "Supposed enchorial alphabet". Moreover, Champollion could not conceivably have missed the fact that Young's rudimentary hieroglyphic 'alphabet' had been derived from the cartouches of Ptolemy and Berenice, as we explained in Chapter 10, "Reading the Rosetta Stone". Surely, having absorbed this 1819 article, and earlier work by Young, Champollion was now primed to take his first correct original step.

It came in January 1822, when he saw a copy of an obelisk inscription sent to the National Institute in Paris by the English collector William Bankes, who had had the obelisk removed from Philae (near Aswan) by Giovanni Belzoni and transported to Bankes's country house in England, where it still stands. The importance of the obelisk was that it was bilingual. The base-block inscription was in Greek, while the column inscription was in hieroglyphic script. This, however, did not make it a true bilingual, a second Rosetta Stone, because the two inscriptions did not match. Notwithstanding, in 1818, Bankes realized that in the Greek letters the names of Ptolemy and Cleopatra, Ptolemaic queen, were mentioned, while in the hieroglyphs two (and only two) cartouches occurred—presumably representing the same two names as written in Greek on the base. One of these cartouches was almost the same as a longer cartouche on the Rosetta Stone identified as Ptolemy by Young:

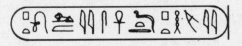

Rosetta Stone

Philae Obelisk

—so the second obelisk cartouche was likely to read Cleopatra. In sending a copy of the inscription to scholars, including the National Institute, Bankes penciled his identification of Cleopatra in the margin of the copy.

Unfortunately for Young, the copy that came to him contained a significant error. The copyist had expressed the first letter of Cleopatra's name with the sign for a T instead of a K. So, says Young, "as I had not leisure at the time to enter into a very minute comparison of the name with other authorities"—this was the period when he took over the *Nautical Almanac*—"I suffered myself to be discouraged with respect to the application of my alphabet to its analysis". In other words, Young had an unlucky break here, but he was also undermined by his lifelong tendency to spread himself.

Champollion, however, was not a man to be diverted from study of Egypt by other interests and duties. He took the new clue—without any acknowledgment to Bankes or Young—and ran with it. Just as Young had done, he decided that a shorter version of the Ptolemy cartouche on the Rosetta Stone spelt only Ptolemy's name:

while the longer cartouche must involve some royal title, tacked on to Ptolemy's name. Again as Young had done, Champollion assumed that Ptolemy was spelt alphabetically, and thus, following Bankes's identification, that the same applied to Cleopatra on the obelisk from Philae. He proceeded to guess the phonetic values of the hieroglyphs in both cartouches:

C	△	P	□
L	🐆	T	◠
E	⧘	O	⧂
O	⧂	L	🐆
P	□	M	⊂
A	🦅	E	⦀⦀
T	⬭	S	⧘
R	⬯		
A	🦅		

There were four signs in common, those with the phonetic values L, E, O and P, but the phonetic value T was represented differently in the two names. Champollion deduced correctly that the two signs for T were what is known as *homophones*, that is, different signs with the same sound (compare in English **J**ill and **G**ill, **C**atherine and **K**atherine)—a concept that Young was also aware of.

The real test of the decipherment, however, was whether these new phonetic values, when applied to the cartouches in other inscriptions, would produce sensible names. Champollion tried them in the following cartouche:

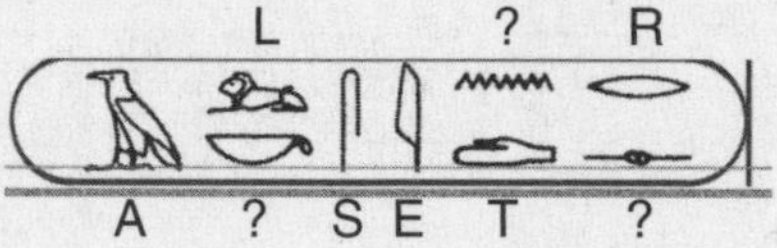

Substitution produced AL?SE?TR?. Champollion guessed ALKSENTRS = (Greek) ALEXANDROS [ALEXANDER]—again the two signs for K/C (𓎡 and 𓈎) are homophonous, as are the two signs for S (𓊃 and 𓋴).

Using the growing alphabet, Champollion went on to identify the cartouches of other rulers of non-Egyptian origin, Berenice (already tackled, though with mistakes, by Young) and Caesar, and a title of the Roman emperor, Autocrator. It was quickly obvious to him that many more identifications would now follow. On 27 September 1822, Champollion felt ready to announce his breakthrough at a meeting of the Academy of Inscriptions, and to follow it in October with the publication of his celebrated *Lettre à M. Dacier*—Bon-Joseph Dacier was the secretary of the academy—in which he unveiled his first shot at a complete hieroglyphic/demotic list of signs with their Greek equivalents, accompanied by a light-hearted cartouche of his own name written in demotic script. (This understandable flourish, which Champollion omitted from his later, more dignified publications, is something not easy to imagine from the pen of his more soberly scientific rival.)

Young was in Paris again at the time and was present at the meeting on 27 September. In fact, he was invited to sit next to Champollion while he read out his paper. It was the first personal encounter of the two decipherers, who were formally introduced by Arago after the meeting, and naturally they had much to discuss, although Young could hardly avoid noticing Champollion's lack of open acknowledgment of his own work. He wrote to Hudson Gurney from Paris:

> Fresnel, a young mathematician of the Civil Engineers, has really been doing some good things in the extension and application of my theory of light, and Champollion ... has been working still harder upon the Egyptian characters. He devotes his whole time to the pursuit and he has been wonderfully successful in some of the documents that he has obtained—but he appears to me to go too fast—and he makes up his mind in many cases where I should think it safer to doubt. But it is better to do too much than to do nothing at all, and others may separate the wheat from the chaff when his harvest is complete. How far he will acknowledge everything which he has either borrowed or might have borrowed from me I am not quite confident, but the world will be sure to remark *que c'est le premier pas qui coûte*, though the proverb is less true in this case than in most, for here every step is laborious. I have many things I should like to show Champollion in England, but I fear his means of locomotion are extremely limited, and I have no chance of being able to augment them.

Young's work was conspicuously downplayed in the *Lettre à M. Dacier*—so patently, in fact, that anyone knowledgeable of the recent history of the Rosetta Stone could not fail to conclude that Champollion had done this deliberately. As Young would remark publicly the following year, with notable understatement: "I did certainly expect to find the chronology of my own researches a little more distinctly stated." Champollion's first publication of the decipherment shows that from the very beginning he was set on keeping all the glory for himself, since he could have had no other motive to downplay Young's role in October 1822, before Young had made a single public criticism of him or his work.

His attitude to Young comes out most clearly if we consider Champollion's description of how Cleopatra's cartouche was identified and used to construct an alphabet, as translated by Young himself from the *Lettre à M. Dacier* (the italic emphases are also Young's):

The hieroglyphical text of the inscription of Rosetta exhibited, on account of its fractures, *only the name of Ptolemy*. The obelisk found in the Isle of Philae, and lately removed to London, contains also the hieroglyphical name of *one of the Ptolemies*, expressed by the same characters that occur in the inscription of Rosetta, surrounded by a ring or border, which must necessarily contain the proper name of a woman, and of a queen of the family of the Lagidae, since this group is terminated by the hieroglyphics expressive of the *feminine* gender; characters which are found at the end of the names of all the Egyptian goddesses without exception. The obelisk was fixed, it is said, to a basis bearing a Greek inscription, which is a petition of the priests of Isis at Philae, addressed to King Ptolemy, to Cleopatra his sister, and to Cleopatra his wife. Now, if this obelisk, and the hiero-glyphical inscription engraved on it, were the result of this petition, which in fact adverts to the consecration of a monument of the kind, the border, with the feminine proper name, can only be that of one of the Cleopatras. This name, and that of Ptolemy, which in the Greek have several letters in common, were capable of being employed for a comparison of the hieroglyphical characters composing them; and if the similar characters in these names expressed in both the same sounds, it followed that their nature must be entirely phonetic.

There is not even a nod here to Young (or Bankes). The fact stung him—encouraged by his friend Gurney—into publishing a book for a general readership, this time under his own name, entitled *An Account of Some Recent Discoveries in Hieroglyphical Literature and Egyptian Antiquities*. He comments on the previous passage by Champollion as follows:

This course of investigation appears, indeed, to be so simple and so natural, that the reader must naturally be inclined to forget that any preliminary steps were required: and to take it for granted, either that it had long been known and admitted, that the rings on the pillar of Rosetta contained the name of Ptolemy, and that the semicircle and the oval constituted the female termination, or that Mr Champollion himself had been the author of these discoveries.

It had, however, been one of the greatest difficulties attending the translation of the hieroglyphics of Rosetta, to explain how the groups within the rings, which varied considerably in different parts of the

pillar, and which occurred in several places where there was no corresponding name in the Greek, while they were not to be found in
others where they ought to have appeared, could possibly represent
the name of Ptolemy; and it was not without considerable labor that
I had been able to overcome this difficulty. The interpretation of the
female termination had never, I believe, been suspected by any but
myself: nor had the name of a single god or goddess, out of more
than five hundred that I have collected, been clearly pointed out by
any person.

But, however Mr Champollion may have arrived at his conclusions,
I admit them, with the greatest pleasure and gratitude, not by
any means as superseding my system, but as fully confirming and
extending it.

Indeed, Young added a provocative subtitle to his book: "Including the
Author's Original Alphabet, As Extended by Mr Champollion".

Champollion was duly provoked. In March 1823, having seen only an
advertisement for the new book, he wrote angrily to Young: "I shall never
consent to recognize any other original alphabet than my own, where it is a
matter of the hieroglyphic alphabet properly called; and the unanimous
opinion of scholars on this point will be more and more confirmed by the
public examination of any other claim." Scholarly war had been declared.

Young's supporters felt that he had taken the vital first steps that had
enabled Champollion to advance, and that Champollion had either
ignored these or claimed that he had come to the same conclusions inde-
pendently. "Nothing can exceed the effrontery of Champollion in thus
complaining to Dr Young, the author of the discoveries ... as if he himself
were the person aggrieved," wrote John Leitch, the editor of Young's linguistic works. Champollion's supporters argued, by and large, that while
Young had taken some first steps, not all of them were correct, as witness
his misreading of some of the signs in the cartouches of Ptolemy and
Berenice. Champollion, they said, had established a *system* that worked
easily when applied to new cartouches, as opposed to Young's more ad hoc
methods, that in some cases required ingenious manipulation to produce
phonetic values. And inevitably they pointed to Champollion's truly revolutionary progress from 1823 onwards, which Young himself generally
admired.

At the end of the chapter, "Mr Champollion", in his book, Young summarized his basic wish:

[that] the further [Champollion] advances by the exertion of his own talents and ingenuity, the more easily he will be able to admit, without any exorbitant sacrifice of his fame, the claim that I have advanced to a priority with respect to the first elements of all his researches; and I cannot help thinking that he will ultimately feel it most for his own substantial honor and reputation, to be more anxious to admit the just claims of others than they be to advance them.

This was a reasonable, temperate request, but it fell on stony ground. Either Champollion had too much vanity to concede anything important to Young, or he had genuinely convinced himself, through his long years of obsession with ancient Egypt, that the crucial first steps were really taken by him—or perhaps there was an amalgam of both feelings in his mind. By sticking intransigently to his claim of sole authorship, he achieved his ambition and came to enjoy general acceptance as *the* decipherer of the Egyptian hieroglyphs. But in so doing he lost his good name. Young was right in his gentle warning: Champollion's personal reputation will forever be tainted by his hubris toward Young.

By the time that Young's book appeared, Champollion was already far ahead of him in the hieroglyphic decipherment. In April 1823, he unveiled his second great breakthrough, which he had hinted at in his *Lettre à M. Dacier*. There he had shown that his alphabet could be applied to the cartouches of ancient rulers of *Egyptian* origin, the pharaohs, as well as to the more recent non-Egyptian Ptolemies of Greek and Roman times. In particular, he had identified a cartouche from Abu Simbel that seemed to spell the name Ramses, a king who, according to a well-known Greek history of Egypt written by the Ptolemaic historian Manetho in the third century BC, belonged to the nineteenth dynasty of ancient Egypt. Now, in his second publication six months after the *Lettre*, Champollion successfully began to apply his alphabet to the main text in the hieroglyphic script—not just the royal names in the cartouches. From this, he found the courage to reject and transcend the centuries-old, stifling belief that hieroglyphic was an entirely conceptual script that used phonetic signs only to represent non-Egyptian names. With this radical new assumption—that the writing system of the ancient Egyptians, both the hieratic/demotic *and* the hieroglyphic, was a mixture of conceptual signs and phonetic signs—Champollion was able to transliterate hundreds of ordinary hieroglyphic words. In many cases, he knew that his transliteration was likely to be correct because it resembled a word in Coptic with a meaning that made sense in the hieroglyphic context (Coptic being the most recent stage of the ancient Egyptian language). It is mainly the Coptic clue that enables us to guess roughly how the hieroglyphic inscriptions must have sounded when read aloud.

In 1824, after many more months of intensive study of hieroglyphs in various Egyptian inscriptions and papyrus manuscripts, Champollion published his definitive statement of his decipherment, *Precis du Système Hiéroglyphique des Anciens Égyptiens.* In his introduction, he made a point of stating what he saw as Young's contribution:

> I recognize that he was the first to publish some correct ideas about the ancient writings of Egypt, that he also was the first to establish some correct distinctions concerning the general nature of these writings, by determining, through a substantial comparison of texts, the value of several groups of characters. I even recognize that he published before me his ideas on the possibility of the existence of several sound-signs, which would have been used to write foreign proper names in Egypt in hieroglyphs; finally that M. Young was also the first to try, but without complete success, to give a phonetic value to the hieroglyphs making up the two names Ptolemy and Berenice.

Perhaps it is superfluous to comment much further. Champollion's statement, though not inaccurate, is clearly grudging and damns Young with faint praise in its vague references to "correct ideas" and "correct distinctions". It fails to articulate Young's two key perceptions of general principles, published in 1819: first, that the demotic (enchorial) script to some extent resembled the hieroglyphic script visually and hence that the former script was derived from the latter; second, that the demotic script was therefore not an alphabet but a mixture of phonetic signs and hieroglyphic signs. This line of argument was what led Young to suggest that the hieroglyphic script too might contain some phonetic elements (for spelling non-Egyptian names like Ptolemy), more than two years before his rival.

Young's own mild verdict on the conflict, as stated in his autobio-graphical sketch, was: "He found that it is easier to gain credit in England for literature than for science; while he observed that, on the continent, there was more candor and indulgence among men of science than among scholars." Though Young does not specifically say so, one can hardly doubt that he was alluding here to the different receptions in England and France accorded to his wave theory of light and his work on the Rosetta Stone. In France, Fresnel, the scientist, had done full and prompt justice to Young's work; Champollion, the scholar, had done it a tardy injustice.

After 1823, Young did not contribute much to the hieroglyphic deci-pherment, and in 1827 he abandoned his work altogether. But he certainly did not abandon the writing of ancient Egypt. Instead, he turned from the hieroglyphic to the demotic script, inspired by an amazing accidental encounter with a papyrus manuscript in late 1822 that we shall describe in the next chapter. This fluke—which Young wrote about at length in *An Account of Some Recent Discoveries in Hieroglyphical Literature and Egyptian Antiquities*—stimulated a late flowering of Egyptian activity in Young. Reading demotic became one of the manifold activities that would fill the final years of this indefatigable polymath.

Chapter 16

A Universal Man

"He might for example, have been styled without impropriety and almost with equal justice, in the middle of a history of his life, a physician, a classical scholar, a linguist, an antiquarian, a biographer, an optician, or a mathematician ... Whether the public would have been more benefitted by his confining his exertions within narrower limits, is a question of great doubt."

Young, "Autobiographical sketch", 1826/27

Apart from deciphering the demotic script, and publishing scientific papers in areas comparatively new to him—atmospheric refraction, the density and shape of the earth, and the theory of life insurance—in the 1820s, Young also became a well-paid "inspector of calculations" and physician to a newly founded society for life insurance. At the same time, he continued to be secretary of the Board of Longitude and superintendent of the *Nautical Almanac*, a physician at St George's Hospital, an active member of the Royal Society and its long-time foreign secretary, and a leading intellectual figure in London society. Bearing in mind the entire spectrum of his earlier work and achievements since 1800, he deserves to be called a Renaissance man or *uomo universale*, like Goethe, Benjamin Franklin or Young's friend Alexander von Humboldt (to whom he dedicated his 1823 book on Egypt)—even, maybe, the most eminent example of such a man in his age.

The advantages of his unique position, and the disadvantages, clearly preoccupied and disturbed Young. For in his autobiographical sketch, which was written during this period, he gives his own ambivalent view of polymathy at considerable length, while remaining modestly silent on several of his major achievements. Part of what he wrote is quoted at the opening of this chapter; Young then continues:

> [H]is own idea was, that the faculties are more exercised, and therefore probably more fortified, by going a little beyond the rudiments only, and overcoming the great elementary difficulties, of a variety of studies, than by spending the same number of hours in any one pursuit: and it was generally more his object to cultivate his own mind than to acquire knowledge for others in departments which were not his immediate concern: while he thought with regard to the modern doctrines, of the division of labor, that they applied much less to mind than to matter, and that while they increased the produce of a workman's physical strength, they tended to reduce his dignity in the scale of existence from a reasoning being, to a mere machine.

Then—still of course referring to himself in the third person—he makes a perceptive remark about the way in which science progresses (which incidentally by implication suggests why today's government funding of scientific research can never be straightforwardly tied to useful applications):

> It is indeed so impossible to foresee the capabilities of improvement in any science, that it is idle to form any general opinion of what would be the comparative advantage of the employment of time in any one investigation rather than another, for almost all the authors of important discoveries and even of inventions, are led as much by accident as by system to their successes. He would probably not have recommended the plan of his own studies as a model for the imitation of others: and he certainly thought that many hours, and even years of his life, had been occupied in pursuits that were comparatively unprofitable. But it is probably best for mankind that the researches of some investigators should be conceived within a narrow compass, while others pass more rapidly through a more extensive sphere of research.

Young's appointment in life insurance came about during a speculative financial boom in 1824-25 that saw 624 life insurance schemes projected, of which no more than a fifth survived their infancy. The Palladium

Life Insurance Company was one of them; it amalgamated in 1856 with the Eagle, and this company later became Eagle Star Assurance. Young was appointed the company's inspector of calculations and its physician in March 1824 at a salary of £500 per annum, making his overall income, including his salary from the Admiralty, his medical practice and his writings, "adequate to his utmost wishes, without any further dependence on the caprice of public opinion in a medical capacity" (the autobiographical sketch again). Later, this salary was reduced to £400 at Young's own suggestion, after he had ascertained that the true amount of work was less than he had expected. "A rare example of conscientiousness in the administration of such institutions, which are not infrequently less designed by their founders for the benefit of the general body of shareholders and insurers, than of the officers who conduct them"—as remarked only too accurately by an acid George Peacock, who was clearly no admirer of the actuarial profession.

One presumes that Young was approached by the Palladium in the first place because in 1816 he had published an anonymous paper entitled "An algebraical expression for the value of lives". This probably grew out of his interest as a physician in the effect of climate on mortality. In his book on consumptive diseases, he included a table of the annual mortality in the different counties of Great Britain, based on the census returns of 1811. The county of Middlesex, which contains London, came off worst, with an annual mortality of 1 in 36, that of Northamptonshire came about half way down the table at 1 in 52, while the rural counties of Wales were the healthiest, at 1 in 73, that is, half the mortality rate of London. "It is obvious that those counties, which contain large manufacturing towns, exhibit a mortality wholly independent of their climate," observed Young, "while the natural salubrity of others, for instance, Cornwall [1 in 62], is probably rendered more conspicuous by their exemption from sedentary employments." Obviously Young was already aware of the complexity of mortality statistics. When he became professionally involved with life insurance in the 1820s, he wrote five more articles under his own name directed at three basic ends: to obtain formulas that could be applied by actuaries in any part of the country, to fit these formulas to certain existing tables of mortality, and to criticize certain actuaries and societies.

Life insurance is a rebarbative subject for most people. Although Young's papers on it do not merit the 'pioneering' label that adheres naturally to his work in physics, physiology and Egyptology, it is worth looking briefly at his dispute, in 1826 and after, with one particular life insurance expert. William Morgan, the chief actuary from as far back as 1775 of the well-established Equitable Society and a fellow of the Royal Society, was known for defending the Equitable's use for the whole country of the Northampton table of mortality, drawn up by Morgan's uncle, Richard Price, a founder of both the theory and practice of life insurance, from observations of mortality in Northamptonshire in the years 1735-80. On this basis, Morgan had managed the Equitable "with greater prudence than equity" (Peacock again) and ensured a flow of profits to its members for more than half a century. But Young was far from convinced that the Equitable was being equitable in applying the Northampton table everywhere—to London, say, and other major cities—especially as the society was still employing Price's original assumption that the population of the country was static, which it most definitely was not by the 1820s. (All formulas for calculating life insurance must take account of both mortality rates and birth rates.)

Young therefore attacked Morgan in a paper for the Royal Society entitled "A formula for expressing the decrement of human life", which concluded:

> I sincerely hope that these considerations may help to undeceive the too credulous public, who have of late not only received some hints that tend to insinuate the probability of an occasional recurrence of a patriarchal longevity, but who have been required to believe, upon the authority of a most respectable mathematician, that the true and unerring value of life is not to be obtained by taking an average of various decrements, but by adopting the extreme of all conceivable estimates, founded only on a hasty assertion of Mr Morgan, and unsupported by any detailed report; an estimate which makes the great climacteric of mankind [i.e., the years in which the greatest number of adults die] in this country, not a paltry 54, or the too much dreaded 63, but no less than EIGHTY-TWO! An age to which nearly one sixth of the survivors at ten are supposed to attain!

An irritated Morgan not surprisingly responded in stout defense of himself, the late Dr Price, his Northampton table and the Equitable Society:

> The public have lately been overwhelmed with tables of the decrements of human life, formed either by amalgamating all the old tables into one heterogeneous mass, and thus giving the true probabilities of life in no place whatever, or by interpolating some of the decrements in one table into those of another; for which purpose a vast variety has been given of complicated and useless formulas. But little or no advance has been made in determining more correctly the probabilities and duration of human life. The tables published in the Report of the Committee of the House of Commons are in general so incorrect, and some of them are even so absurd, as to be unfit for use; and serve only to encourage the popular delusion of the improved healthiness and greater longevity of the people of this kingdom.

Young was correct to question Morgan's outdated and self-interested methods of calculation, but his own empirical formula, with some twenty constants, was so complicated that it was altogether impracticable for the calculation of annuities. Although Young's principles of life insurance merit a footnote in the history of the subject, in practice they exerted no perceptible influence on the development of the life insurance industry.

More fruitful, if rather less lucrative, was his simultaneous scientific work in geology and geodesy. Young had long been interested in this subject; indeed he was among the very first to understand an important aspect of earthquakes: the similarity of the vibrations caused by earthquakes to the longitudinal vibration of sound waves. In his *Natural Philosophy*, Young states that "where the agitation produced by an earthquake extends further than there is any reason to suspect a subterraneous commotion, it is probably propagated through the earth nearly in the same manner as a noise is conveyed through the air." Now, in the 1820s, he turned his attention to the long-debated density and 'figure' of the earth, that is, "the shape assumed by a self-gravitating, rotating mass of fluid" (in the words of a current mathematician who has studied the debate, Alex Craik). Newton had maintained that the spinning earth was not a sphere, but a spheroid slightly flattened at the poles and slightly bulging at the equator, and after some decades of controversy about whether the flattening and bulging were in fact the other way around, two grueling scientific expeditions set out from France in the mid-eighteenth century to conduct trigonometrical surveys in Lapland (near the pole) and Peru (at the equator), and eventually proved that Newton was correct. By Young's time, however, it had become clear that the earth was not an exact spheroid and that further theoretical refinements to its figure were necessary to take account of the fact that the planet was not of uniform density.

Pierre-Simon Laplace was interested in the problem, too. Despite his disagreement with Young on his wave theory of light and other scientific matters, Laplace was impressed by one of Young's arguments and adopted it in his own work. "Until now," he wrote, "mathematicians have not included in this research the effect resulting from the compression of the strata. Dr Young has called their attention to this object, by the ingenious remark, that we may thus explain the increase of density of the strata of the terrestrial spheroid." Nevertheless, Young disagreed with an assumption of Laplace, that the elasticity of a solid must be proportional *not* simply to its density, as was known to be true of elastic fluids, but to the square of its density. "M. Laplace's hypothesis is not correctly applicable to the internal structure of the earth; since it either makes the mean density too small in comparison with that of the surface, or the compressibility at the surface too great ... In this respect the simple analogy of elastic fluids will afford us a result more conformable to observation." Young proceeded to show that with the assumption of simple proportionality and with a modulus of elasticity for rock of ten million feet, the figure of the earth that emerged was one close to that actually observed.

A new method of calculating the figure of the earth "from a single tangent" was among Young's last scientific calculations, found among his papers after his death. As he told Hudson Gurney at the time: "it is my pride and pleasure as far as I am able to supersede the necessity of experiments and especially of expensive ones. I have just been inventing a mode of determining the figure of the earth from two points in sight of each other, without going either to Lapland or to Peru".

Egyptian writing also continued to absorb him to the very end, as we know. Young may have let slip the hieroglyphic crown to Champollion, but the prize for deciphering demotic—what he called enchorial—was still available. And this time he felt that luck was with him.

In 1821, when Young was in Italy, he had tried desperately hard to acquire a copy of Drovetti's bilingual inscription, mainly so as to aid and confirm his own interpretation of the demotic portion of the Rosetta Stone. Then, by a stroke of great good fortune, one day in November 1822 he was lent a box of papyri by George Francis Grey, a friend of an old Cambridge University friend, who had bought them from an Arab at Thebes. That very evening Young discovered to his absolute astonishment that two of Grey's papyri contained a Greek translation of a demotic papyrus from a totally different source that Young had been trying to decipher without much success. Miraculously, he now had a real bilingual in his hand, and could forget all about how to inveigle a copy of Drovetti's inscription out of its jealously proprietorial owner. The candles in No. 48 Welbeck Street must have burnt until dawn on the night of 22-23 November 1822. A few months later, Young memorably described his almost feverish excitement at the find in his *Account of Some Recent Discoveries in Hieroglyphical Literature and Egyptian Antiquities*:

> I could not, therefore, but conclude, that a most extraordinary chance had brought into my possession a document which was not very likely, in the first place, ever to have existed, still less to have been preserved uninjured, for my information, through a period of near two thousand years: but that this very extraordinary translation should have been brought safely to Europe, to England, and to me, at the very moment when it was most of all desirable to me to possess it, as the illustration of an original which I was then studying, but without any other reasonable hope of being able fully to comprehend it; this combination would, in other times, have been considered as affording ample evidence of my having become an Egyptian sorcerer.

It was an inspiring moment, and Young made solid progress with demotic over the next few years, as Champollion raced ahead with the hieroglyphs (while also himself studying demotic). But being Young, he got diverted—by annuities, the figure of the earth, the *Nautical Almanac* and many other appealing byways of knowledge. Then, in June 1827, he received a letter in Latin that seems to have galvanized him again. It was written by Amedeo Peyron, an Italian specialist in Coptic at Turin (the place where, by chance, Drovetti's elusive stone now rested), and it mixed high praise of Young with some tactful criticism:

You write that from time to time you will publish new material which will increase our knowledge of Egyptian matters. I am very glad to hear this and I urge you to keep your word. For, as Champollion will witness, and other friends to whom I have mentioned your name, I have always felt and so do many others, that you are a man of rare and superhuman genius with a quick and penetrating vision, and you have the power to surpass not only myself but all the philologists of Europe, so that there is universal regret that your versatility is so widely engaged in the sciences—medicine, astronomy, analysis, etc. etc. that you are unable to press on with your discoveries and bring them to that pitch of perfection which we have the right to expect from a man of your conspicuous talents; for you are constantly being drawn from one science to another, you have to turn your attention from mathematics to Greek philosophy and from that to medicine etc. The result is that there are some mistakes in your books which you yourself might well have corrected.

From now until his death two years later, Young worked assiduously at his *Rudiments of an Egyptian Dictionary in the Ancient Enchorial Character; Containing All the Words of Which the Sense Has Been Ascertained.* And it is pleasant to record that Champollion, who was now the curator of the Egyptian collection at the Louvre Museum in Paris, assisted him. In the summer of 1828, Young visited Paris to accept his recent honor of being elected as one of the eight foreign associates of the National Institute, just before Champollion set off for Egypt. Young told Gurney that Champollion "has shown me far more attention than I ever showed or could show, to any living being: he devoted *seven* whole hours at once to looking over with me his papers and the magnificent collection which is committed to his care ... he is to let me have the use ... of all his collections and his notes relating to the enchorial character that I may make what use I please of them." We can only guess at Champollion's motives: no doubt they included some new respect for Young as a foreign associate of the National Institute, but more important must have been Champollion's pride in his invulnerable achievement and in his curatorship; plus, it would surely be reasonable to assume some feeling of guilt at his unacknowledged debt to Young. Anyway, Young was careful to acknowledge Champollion's help in generous terms in his dictionary. Although the

difficulties of deciphering enchorial/demotic remained formidable—
many manuscripts contain puzzling passages even today—Young could
justifiably claim that "thirty years ago, not a single article of the list [of
words in the dictionary] existed even in the imagination of the wildest
enthusiast: and that within these ten years, a single date only was tolerably
ascertained, out of about fifty which are here interpreted, and in many
instances ascertained with astronomical precision." The Egyptologist John
Ray sums up: "Young was the first person since the end of the Roman
Empire to be able to read a demotic text, and, in spite of a proportion of
incorrect guesses, he surely deserves to be known as the decipherer of
demotic. It is no disservice to Champollion to allow him this distinction."

*"Far more attention than I ever showed or could show, to any living
being"*—it is a faintly shocking remark from Young to his oldest friend,
especially since it refers to his intellectual sparring partner Champollion.
The remark appears in Alex Wood's biography, but not in the otherwise
identical quotation from Young's letter in Peacock's book, where these few
words are simply missing. This may have been due to a copying error by
Peacock; however, it seems at least possible, and even probable, that
Peacock deliberately omitted the unguarded remark out of delicacy for the
feelings of Young's wife Eliza. No doubt, after a quarter of a century's marriage, she was profoundly aware of her husband's absorption in the pursuit
of knowledge, but this remark seems to imply an emotional detachment
from other human beings verging on the inhuman.

Perhaps such detachment in a leading scientist is hardly news. The
popular caricature of scientific genius today is generally somewhat misan-
thropic. And in reality, Newton, notoriously, and his polymathic contemporary Hooke, and also Einstein, all sacrificed intimate personal
relationships for scientific insights. However, Einstein could write, aged 70,
in a memorial message for a Jewish friend: "Knowledge exists in two
forms—lifeless, stored in books, and alive in the consciousness of men.
The second form of existence is after all the essential one; the first, indis-
pensable as it may be, occupies only an inferior position." Young would not
have agreed to relegate books in this way, and his emphasis, one feels,
would have been on the cultivation and perfection of the consciousness of
one man, oneself, rather than the sharing of one's knowledge with others.
(Hence his lack of success as a lecturer and, possibly, as a physician.) Young

was a man who probably felt most alive at times of solitary reflection in his study. Another letter to Gurney, written in 1820, catches this mood well: "I have derived more pleasure within these few days from a contemptuous hint of a great mathematician, which I can at once show to be unjust, and from an elaborate attempt to substitute a new theory for one of mine, which I can easily prove to be far less accurate, than I should probably have received from the most fulsome compliments".

Yet—and this is what makes Young an intriguing person in addition to being a fascinating thinker—he was genuinely sensitive to the arts and could often be distinctly fond of human company. Writing to his favorite sister-in-law Emily from Worthing, he says: "I have been dashing through vocal and instrumental music without any reserve or modesty, being determined to keep myself in practice for the pleasure of accompanying you." In another letter to her, he writes of a small dinner party—"one of the very few dinners that reconcile one to living in London"—at Lord Elgin's, with only the American-born historical painter Benjamin West, president of the Royal Academy, and a couple of others present. Young describes West for her:

[He] is really a most interesting personage in everything that relates to his profession; in other respects he is very much like any other man of 74: but he was only seven months painting his great picture [*The Death of General Wolfe*], which he sold for 3000 guineas, and which produced 13,000 to the British Institution by its exhibition. His present picture is visited daily by 472 people on an average of a fortnight. He paints fifteen hours a day, not requiring any other exercise, and sleeping but seven: he paints without any model, in order to avoid introducing portraits, and to preserve an ideal character of perfection in his figures; but when he has once drawn them, he corrects the attitudes and the lights by comparison with a real figure: for this reason, he said, he never repeated himself. To me it appeared that he did very often repeat the same kind of countenance, and his mode of painting seemed to explain the reason of it, and I ventured to hint something of the kind in an indirect manner. In consequence of his wishing to see me at his house, I called on him last Sunday, and sat a long while with him. He perfectly remembered my once having seen him 21 years ago, when, as he observed, I was dressed in a different costume [as a Quaker]. He told me the history of the little Cupid and Psyche which I have [one of the paintings bequeathed to Young by his great-uncle Richard Brocklesby]: he painted it in the year 1760, when he was in his twenty-first year, before he had ever been in England.

After a lot more description of his Sunday spent with West, Young signs off insouciantly: "And now I have told you enough of Mr West, a man who has covered 7000 square feet of canvas, and too much, except that I think all these particulars worth remembering, and therefore worth writing; and if you do not think them worth reading, you are at liberty to pass them over and burn the letter."

Boswell on Doctor Johnson this account may not be, but it is scarcely the writing of an uncongenial scientific recluse. Nor does Young's own portrait, painted from life some time after 1822 at the request of Hudson Gurney by Sir Thomas Lawrence, the leading portrait painter of his time, celebrated for capturing a true likeness of his subjects, suggest unworldliness or misanthropy. Even after making allowance for an artist's desire to please, the dominant impression from the portrait is one of intelligence and determination, but also sensitivity and openness to the world.

All of these qualities were more than evident in Young's relations with the Royal Society over a period of more than thirty years. In fact, Young epitomizes the Royal Society ethos at its finest. Yet he never once made a speech at a council meeting, and when, in 1827, following his election as a foreign associate of the National Institute of France, he was mentioned as a possible president of the Royal Society, after Sir Humphry Davy had to retire through ill health, he demurred. "I find there has been pretty general conversation about making *ME* president of the Royal Society," he wrote to Emily, "and I really think if I were foolish enough to wish for the office, I am at this *moment* popular enough to obtain it; but you know that nothing is farther from my wishes." Instead, Davies Gilbert, politician and promoter of science but no scientist himself, was elected. Young liked Gilbert, but doubted his capacity to control unruly council meetings. "I told him that he had not quite enough of the devil in him; that Sir Joseph Banks should have left his *eyebrows* to go with his cocked hat, if he left the society nothing else."

By now, Young was certainly an established figure—both in science and in life. At the end of 1825, he left Welbeck Street after 25 years there, and moved a mere half a mile to a grand new house at 9 Park Square in Regent's Park, just north of John Nash's imposing, stuccoed development of Regent Street. Probably his income from the Palladium Life Insurance Company went into building this residence. At Park Square, to quote Gurney's memoir, "he led the life of a philosopher, surrounded by every domestic comfort, and enjoying the pleasures of an extensive and culti-vated society, who knew how to appreciate him." That meant, to use Young's own words, "the pursuit of such fame as he valued, or of such acquirements as he might think to deserve it."

The note of complacency, though readily understandable, is unmis-takable, and it is bound to provoke a reaction from a more egalitarian age. According to Geoffrey Cantor, writing in the latest *Dictionary of National Biography*, "Living in a period when the social elite was under attack both at home and abroad, Young never wavered in his defense of the status quo and he remained mindful of his position as an English gentleman." True enough: but did this make Young an Establishment figure, as implied by Cantor's comment? I think not. He started his career with a considerable inheritance, but the rest of his money was self-made, from his own untir-ing work as a scientist, mathematician, physician and writer. Intellectually and socially he was seldom easily or totally accepted, unlike, say, his equally middle-class but ambitious and fashionable contemporary Davy. When Young died, he was still plain Dr Young, signally lacking in the national honors normally awarded to eminent scientists and public servants. If he yearned to be Sir Thomas, like Sir Humphry Davy, he practiced few of the usual flatteries of the rich and powerful required to secure a knighthood. Most probably, had a title been offered to him on his own terms, Young would have been pleased to accept it; but he was not willing to compro-mise his principles to obtain such a symbol of popularity—whether in his Royal Institution lectures, his practice of medicine or his superintendence of the *Nautical Almanac*. Establishment figures are comfortable with com-promise and become fixed in their principles only with prosperity and age. With Young, "As far as the qualities of the mind and feelings are concerned, he may be said to have been born old, and to have died young"—to repeat an earlier comment he made on himself in his autobiographical sketch.

Death came relatively swiftly to him. After a lifetime in which he had not been confined to bed for a single day, even during his adolescent brush with incipient consumption, he experienced unusual fatigue while visiting Geneva in the summer of 1828. Then, in February 1829, he suffered what he apparently considered to be repeated attacks of asthma, and at the beginning of April he had great difficulty in breathing, with some discharge of blood from the lungs and great weakness. But he continued to work, eventually from his bed, and to arrange his affairs through Gurney. The attacks on him by astronomers intent on gaining control of the *Nautical Almanac*, after the abolition of the Board of Longitude in 1828, were at their peak. Young had earlier replied in forceful detail, but now he declined to respond further and asked Gurney "that nothing should go forth on his part to increase irritation". Instead he worked steadily, if feebly, on the last stages of correcting the proofs of his *Rudiments of an Egyptian Dictionary*. There are a few proof pages in Young's manuscripts at the British Library, and it is moving to see his precise handwritten corrections in red ink; at the very end, unable to hold a pen, he was reduced to working in pencil. He told Gurney: "that it was a work which if he should live it would be a satisfaction to him to have finished, but that if it were otherwise, which seemed most probable, as he had never witnessed a complaint which appeared to make more rapid progress, it would still be a great satisfaction to him never to have spent an idle day in his life." In the event, he reached page 96 of the proofs, almost to the end of the book, before expiring on the morning of 10 May, just short of 56 years old.

The post-mortem examination carried out by Young's St George's Hospital colleague Benjamin Brodie on the following day revealed no tubercular damage to the lungs. However:

> The parietes of the heart but especially those of the left ventricle were of unusual thickness (the latter, might be of double the usual thickness). The sigmoid valves of the aorta were very slightly ossified in spots. The aorta from its origin to its bifurcation was ossified to a very great extent, so as to form throughout the greater part of its extent a hard and unyielding tube. It had also lost its cylindrical form: bulging out in some parts, contracted and indented in other parts, and altogether considerably diminished in diameter.

No doubt Young himself, who had made detailed hydraulic calculations on the circulation of blood in the heart in his Royal Society lecture of
1808, would have been fascinated. He was a relatively frugal eater, who neither smoked nor drank alcohol, and a regular taker of exercise. Neither of
his parents had died young. What could be the cause of such extensive ossification of his aorta in his mid-fifties? His friend Gurney put it down to
Young's "unwearied and incessant labor of the mind from the earliest days
of infancy." But a current consultant cardiologist, David Sprigings, thinks
this explanation unlikely to be true:

> "Ossification of the aorta" is what would be recognized today as
> advanced atherosclerosis with calcification. Atherosclerosis of the
> aorta may involve the origins of the arteries to the kidneys, which
> arise from the aorta in the upper part of the abdomen. Severe nar-
> rowing of these arteries can result in kidney failure, high blood pres-
> sure and congestion of the lungs. Progressive and ultimately fatal
> kidney failure, complicated by episodes of severe pulmonary conges-
> tion, would explain the decline in Young's health over the last months
> of his life, and the attacks of acute breathlessness (misdiagnosed as
> asthma).
>
> It is unclear why Young should have such severe atherosclerosis in
> middle age. The major risk-factors for this condition are diabetes
> (there is no evidence that Young had this disorder), high blood pres-
> sure (not known—the sphygmomanometer was not invented until
> later in the nineteenth century), tobacco smoking (he did not smoke)
> and high blood cholesterol. It is possible that Young had a metabolic
> disorder resulting in high blood cholesterol, although we have no evi-
> dence that this was familial; Young's parents lived to good ages for
> their time. The disorder would have developed over several decades.
> While mental stress may be a factor in the clinical manifestations of
> coronary artery atherosclerosis (for example, triggering a heart
> attack), its contribution to the progression of atherosclerosis in the
> aorta and other arteries remains speculative.

The public reaction to Young's death was small. A genuinely shocked
Arago later told the National Institute in his eulogy for their late lamented
foreign associate: "The death of Young in his own country attracted but little regard."

The medical journal, *The Lancet*, carried a brief news item about the death of the "distinguished physician":

> Dr Young, while eminent in his profession was, at the same time, one of the first philosophers in Europe. His readings and researches in natural philosophy were extraordinarily great; the second volume of his works on that subject, displays the extent of his inquiries and acquaintance with the work of other men. Dr Young's name had, of late, been very frequently before the public, through a long controversy between himself and the first astronomers in this country, which was carried on with a degree of acrimony not very befitting philosophers.

The Royal Society, not surprisingly, did better than this. Davies Gilbert, the president, who had known Young fairly well, stated in a valedictory address:

> The multiplied objects which he pursued were carried to such an extent, that each might have been supposed to have exclusively occupied the full powers of his mind; knowledge in the abstract, the most enlarged generalizations, and the most minute and intricate details, were equally affected by him; but he had most pleasure in that which appeared to be most difficult of investigation. ... The example is only to be followed by those of equal capacity and equal perseverance; and rather recommends the concentration of research within the limits of some defined portion of science, than the endeavor to embrace the whole.

There was no other official response. But eventually, at the urging of Mrs Young and the ever-loyal Gurney, space was found in Westminster Abbey for a memorial plaque written by Gurney with a medallion of Young by the sculptor Sir Francis Chantrey. There, in the chapel of St Michael, Young rubs shoulders with Sir James Young Simpson (the discoverer of chloroform), the physician Matthew Baillie (who had taught Young), Sir Humphry Davy (Young's fellow lecturer at the Royal Institution), the 3rd Baron Rayleigh (who greatly admired Young's physics), the engineer Thomas Telford (whose iron bridge proposal Young had supported) and the actress Sarah Siddons (whose performances

Young had watched as a student in Edinburgh). It is suitably diverse and distinguished company for him.

Engraving of the medallion of Young by Sir Francis Chantrey in
Westminster Abbey, as shown in the biography of Young by Peacock.

A year after Young's death, Gurney—while writing his memoir of Young—asked Sir John Herschel for his assessment. Herschel responded at length and then concluded: "how inadequate and limited a view these observations can afford of the extensive scientific labors, and truly original genius of Dr Young. To do anything approaching to justice to his reputation in that respect, would call for the exercise of powers more nearly allied to his own than I can pretend to boast." If this was true then—and with reference only to Young's scientific achievements, not his entire oeuvre—how much truer is it today, almost two centuries later. For those of us lesser mortals who feel instinctively drawn to versatility of genius, Young is guaranteed to be an inspiration; while others whose taste is for genius with a narrow focus (like Fresnel's and Champollion's) will feel bound to regard him with skepticism. What is undeniable, though, is that Thomas Young really did approximate to 'the last man who knew everything'—however much he himself would have denied this—and we can safely say, with the endless expansion and bifurcation of knowledge, that no one will be able to stake this awesome claim ever again.

Notes and References

Since this book is not intended to be a full biography, only references for the quotations in the text are given, with occasional notes. Note that the precise wording of the quotations from Young's letters, the originals of which were available to George Peacock and Alex Wood but have since disappeared, sometimes differs in their two biographies; in each case, I have chosen what appears to me to be the most reliable version.

Preface

ix   **Young probably had** Unsigned note for a Science Museum loan circulation exhibition, 1973 (copy in the J. Z. Young papers at University College London).

Introduction

1   **Fortunate Newton** Newton: vii-viii.

2   **heart of quantum mechanics** Feynman: ch. 37: 2.

3   **anything of a medical nature** Letter to Macvey Napier (12 Feb. 1816) in Peacock: 253.

4   **Hudson's ambition** Caption to a photo of Hudson Gurney and his wife in Anderson.

5   **About this time last year** Letter to Gurney (18 Dec. 1820) in Wood: 317-19. A less-accurate version appears in Peacock: 448-49.

7   **Young occupied a very high place** Rayleigh, "Commemoration lecture": 205-06. The lecture took place on 6 June 1899. When Rayleigh lectured at the Royal Institution in 1892, he used many of the same pieces of demonstration apparatus used by Young in 1802-03 (see Strutt: 234).

7   **[Young] was one of the most acute men** Helmholtz, *Popular Lectures*: 249.

7   **Thomas Young was the first** Quoted in Park: 305.

7   **Cette idée fut** Quoted in Wood: 253.

8   **a man of deep learning** Gardiner: 14.

8   **polymath of amazing reach** Zajonc: 109-10.

8   **Unfortunately for ophthalmology** Fonda: 808.

8   **the founder of all modern neurophysiology** J. Z. Young, "Scientific Autobiography": 1 (unpublished original with his papers in the archive of University College London, probably drafted in 1995/96).

8   **The truth is that, in scientific discovery** Ray, "The name of the first: Thomas Young and the decipherment of Egyptian writing" (unpublished lecture).

9   **Like other members of the Royal Society** Park: 245-47.

9   **Young studied medicine** Cantor, "The changing role of Young's ether": 61.

9   **He was certainly highly intelligent** Cantor, "Thomas Young": 949.

9   **Young was a man with a grievance** Pope: 66-67.

10   **[He did] useful enough work** Ibid: 68.

10   **Even though everything that Champollion said** Ibid: 77.

10   **History is unkind to polymaths** Murray: v.

11   **A Boyle and a Hooke** Young, *Natural Philosophy*, vol. 1: 7.

11   **Hooke was as great** Ibid: 248.

14   **he replied that it was no fatigue** Peacock: 480.

1 Child Prodigy

15 **Although I have readily** Pettigrew: 14.

15 **Biograph: notes whence extracts** Hilts: 248.

16 **wholly without distinction** Westfall: 1.

16 **His father followed** Quoted in Hilts: 250.

16 **of the strictest of a sect** Gurney: 6.

17 **To the bent of these early impressions** Ibid.

17 **His parents were rather below** Quoted in Hilts: 250.

17 **despite the emphasis on discipline** Brooke and Cantor: 296. A list of Quaker and ex-Quaker fellows of the Royal Society on pages 312-13 curiously excludes Richard Brocklesby.

18 **if it was allowable to dwell** Quoted in Hilts: 254.

18 **This poem was repeated** Quoted in Peacock: 3.

19 **though a strong memory** Young, *Miscellaneous Works*, vol. 3: 608.

19 **It seems to be by a wise** Ibid: 620.

19 **As far as the qualities of the mind** Quoted in Hilts: 254.

19 **I like a deep and difficult investigation** Letter to Gurney (no date given) in Wood: 270-71.

19 **He was the best kind of infant prodigy** Asimov: 241.

20 **who had neither talent** Quoted in Hilts: 250.

21 **a very ingenious young man** Quoted in Peacock: 5.

21 **an electrical machine** Ibid: 6.

21 **which became the constant companion...fluxions.** Ibid: 7.

22 **not that he was ever particularly fond** Quoted in Hilts: 251.

22 **[Hunter's] early distrust** Moore, *The Knife Man*: 24.

22 **perhaps the most profitable** Quoted in Hilts: 251.

23 **It is singular** Letter to Gurney (23 Jan. 1816) in Wood: 75.

23   **all sorts of feats** Gurney: 18. The story is told not by Gurney in his *Memoir* but by Dominique Arago, who must have heard it from Gurney himself (see Arago: 231).

24   **a retentive memory** Peacock: 31.

25   ***Pentalogia Graeca*** Quoted in Hilts: 255.

25   **Though he wrote with rapidity** Ibid: 254.

25   **a disease so frequent** Young, *Consumptive Diseases*: 20.

26   **[T]he dust of hard substances** Ibid: 44.

26   **I cannot help being persuaded** Ibid: 53.

26   **twice only** Ibid: 56.

26   **exceedingly painful** Ibid: 68.

26   **the Peruvian bark** Ibid: 74.

26   **little more than water** Ibid: 79.

27   **Not that I am of opinion** Letter to Young (13 Oct. 1789) in Peacock: 15-16.

27   **he was not fourteen** Quoted in Hilts: 254.

28   **I duly received a pleasing letter** Letter to Young (no date given) in Peacock: 19-20.

29   **Young: Will *turba scholarum* do?** Quoted in Peacock: 24.

29   **somewhat characteristically** Young, *Miscellaneous Works*, vol. 3: 614.

29   **British master of classical scholarship** *Encyclopaedia Britannica*, 15th edn: entry for "Porson, Richard".

29   **one of the greatest men** Young, *Miscellaneous Works*, vol. 3: 619.

29   **We find nothing in the nature of theory** Ibid: 614.

31   **[T]he plan for his studying physic** Letter to Barclay (21 Jan. 1791) in Wood: 14-15.

2 Fellow of the Royal Society

33 **It is well known that the eye** Young, *Miscellaneous Works*, vol. 1: 1.

33 **the best spot in Great Britain** Quoted in Porter, *The Greatest Benefit to Mankind*: 293.

34 **In practice this meant** Moore, *The Knife Man*: 45.

34 **Many a bereaved relative** Ibid: 53.

35 **The works of Mr John Hunter** Young, *Consumptive Diseases*: 359.

35 **We have gone through** Quoted in Pettigrew: 5.

36 **Accommodation is the process** Introduction to Young, *Natural Philosophy*, vol. 1: viii.

37 **in closely examining** Young, *Miscellaneous Works*, vol. 1: 4.

37 **I conceive** Ibid: 5.

38 **by no means so clear** Quoted in Peacock: 39.

38 **a gentleman conversant** Ibid: 37.

39 **I hope I am not thoughtless** Letter to Sarah Young (no date given) in Pettigrew: 13.

40 **I really never saw** Letter to Brocklesby (5 May 1794) in Peacock: 42.

40 **a principal reason** Letter to Sarah Young (no date given) in Peacock: 45.

40 **he had predilection** Quoted in Hilts: 252.

3 Itinerant Medical Student

41 **I expect many advantages** Letter to Sarah Young (no date given) in Pettigrew: 6.

42 **I felt his rams** Quoted in Peacock: 48.

43 **Much ingenuity** Young, *Consumptive Diseases*: 364.

43 **He gave me my choice** Quoted in Peacock: 49.

43 **He unites the scholar** Ibid: 49.

43 **He has studied at Cambridge** Ibid: 50.

43   **The university had many attractions** Porter, *The Greatest Benefit to Mankind*: 290-91.

44   **with respect to the study of physic** Letter to Brocklesby (no date given) in Peacock: 52.

45   **this was received with applause** Quoted in Peacock: 53.

45   **with as much respect** Ibid: 54.

46   **There is a namesake of yours here** Letter to Professor Young (no date given) in Dalzel : 118.

46   **I have seen Mrs Siddons** Letter to Bostock (no date given) in Peacock: 59-60.

46   **my friend Cruikshanks** Ibid: 60.

47   **I began, and the next day** Quoted in Peacock: 61.

47   **I was mounted on a stout** Ibid: 63.

47   **was obliged to creep up** Ibid: 65.

48   **To lose one's way** Ibid: 66.

48   **a rich store** Ibid: 67.

48   **passionately fond** Peacock: 60.

48   **read them some of my extracts** Quoted in Peacock: 69.

48   **I was showing Lady C.** Ibid: 72.

49   **after this I must be regulated… virtuous and learned** Letter to Sarah Young (no date given) in Pettigrew: 6.

49   **I am within a hundred yards** Letter to Sarah Young (24 Apr. 1796) in Pettigrew: 7.

50   **all of a sudden two Scotchmen** Letter to Bostock (14 Dec. 1796) in Peacock: 79.

50   **The English physicians are quoted** Letter to Brocklesby (no date given) in Peacock: 87.

51   **The little disposition** Letter to Dalzel (19 Apr. 1797) in Dalzel: 139.

51   **on the whole, one must be** Letter to Brocklesby (no date given) in Peacock: 83.

51   **well furnished with cakes** Quoted in Peacock: 89.

51   **Ricardo Brocklesby** The title of the thesis was *De Corporis Humani Viribus Conservatricibus*. A copy is in the British Library.

51   **married to Hygeia** Peacock: 90.

51   **You say my thesis** Letter to Brocklesby (12 Dec. 1796) in Peacock: 105.

52   **while that great conqueror** Quoted in Hilts: 252.

52   **a little specimen** Quoted in Peacock: 101.

52   **We traveled in the finest moonshine** Ibid: 102.

53   **I have not formed** Letter to Brocklesby (12 Dec. 1796) in Peacock: 105.

53   **[T]here are more learned men** Quoted in Peacock: 110.

4 'Phenomenon' Young

55   **I am at present a good deal** Letter to Dalzel (5 July 1797) in Dalzel: 144.

55   **all persons practicing physic** Quoted in Wood: 54.

56   **bona fide a member** Ibid: 56.

57   **attended places of public diversion** Quoted in Brooke and Cantor: 286.

57   **by his own acknowledgment** Quoted in Cantor, "Thomas Young": 946.

57   **making [a] contribution** Quoted in Brooke and Cantor: 286.

57   **The foolish laws** Letter to Dalzel (19 Apr. 1797) in Dalzel: 138.

57   **he did not think it necessary** Quoted in Hilts: 252.

58   **When I began the outline** Young, *Miscellaneous Works*, vol. 1: 199.

58   **The professors of the university** Quoted in Wood: 57.

58   **stuff Latin and Greek** Quoted in Moore, *The Knife Man*: 99.

58   **I am ashamed to find** Letter to Dalzel (8 July 1798) in Dalzel: 161.

59   **[H]aving learned "to know how little can be known"** Young, *Miscellaneous Works*, vol. 3: 624.

59   **was a man of great energy** Peacock: 115.

60   **When the master introduced…no idle scoff ever escaped him.** Quoted in Peacock: 116-19.

64   **a signal honor** Wood: 61.

64   **that Young does not produce** Quoted in Wood: 61.

65   **Young had just reason** Peacock: 124-25.

65   **enough to afford him** Quoted in Hilts: 252.

5 Physician of Vision

67   **His pursuits** Quoted in Hilts: 253.

68   **for Hercules and Atlas** Quoted in Porter, *London*: 123.

69   **I had now attained** Ibid: 141.

69   **the confusion of furnishing** Letter to Dalzel (27 June 1801) in Dalzel: 206.

70   **masterly monograph** Quoted in Wood: 104. Parsons made the comment in 1930.

71   **In the year 1793** Young, *Miscellaneous Works*, vol. 1: 12-13.

74   **He felt some inconvenience** Quoted in Hilts: 254.

74   **My eye, in a state of relaxation** Young, *Miscellaneous Works*, vol. 1: 26.

75   **that many persons were obliged** Ibid: 26.

75   **For measuring the diameters** Ibid: 25.

76   **With an eye less prominent** Ibid: 25.

76   **I placed two candles** Ibid: 39.

77   **I take out of a small** Ibid: 41.

78   **[A] much more delicate [test]** Ibid: 42.

78   **But no such circumstance** Ibid: 42-43.

78   **obvious advantage** Ibid: 46.

79   **the universal result is** Ibid: 46.

79   **the imperfect eye** Young, *An Introduction to Medical Literature*: 99.

79   **Now, whether we call the lens** Young, *Miscellaneous Works*, vol. 1: 51.

80   **First, the determination** Ibid: 60-61.

80   **Surely the most prescient work** Moore, *Schrödinger*: 122.

81   **Now, as it is almost impossible** Young, *Miscellaneous Works*, vol. 1: 147.

82   **proposed a theory of color vision** Kline: 3. Kline notes that this idea was pointed out to him by J. Z. Young.

82   **the definitive experiments** Hubel: 168.

82   **this has never been observed** "Catalogue—physical optics" in Young, *Natural Philosophy*, vol. 3: 315. See Wade, *A Natural History of Vision*: 136-42, for a brief history of color blindness, which mentions the tests on Dalton's eyes.

6 Royal Institution Lecturer

85   **I shall esteem it better** Young, *Natural Philosophy*, vol. 1: 8.

85   **split London's surgical fraternity** Moore, *The Knife Man*: 242.

85   **Gentlemen...** Ibid: 250.

86   **[M]y colleague ... even in his** Entry in Young's notebooks for his lectures, 17 May 1802, quoted in Cantor, "Thomas Young's lectures at the Royal Institution": 93. The notebooks are kept in the archive of University College London.

86   **never either very popular** Quoted in Hilts: 252.

86   **a narcoleptically boring speaker** Thomas: 20. On the other hand, the Royal Institution director, Sir John Meurig Thomas, describes Young as "a brilliant all-rounder: a physicist, a physician, a physiologist and a philologist."

87   **must even now be held** Bence Jones: 222.

87   **Proposals for forming** Quoted in James: 1.

88   **I am willing to undertake** Letter to Rumford (9 July 1801) in Wood: 119-20.

89   **an immediate repetition** Letter to Dalzel (29 Mar. 1802) in Dalzel: 213.

89   **For example, the first includes** Young, *Natural Philosophy*, vol. 1: vii.

89   **primary and peculiar object...so elegant and so rational.** Ibid: 2-3.

90   **I shall in general entreat** Ibid: 7-8.

90   **A considerable portion** Ibid: 3.

91   **I remember ... his taking me** Quoted in Peacock: 118.

91   **Scientific Researches!** The caricature appears in Godfrey: 204.

92   **the *effect* on him** Entry in Lady Holland's diary for 22 Mar. 1800 quoted in Cantor, "Thomas Young's lectures at the Royal Institution": 95. On page 110 of this article, Geoffrey Cantor discusses the identification of the first lecturer and prefers Garnett to Young. Other scholars tend to prefer Young, for example Godfrey in his catalog of Gillray drawings, and the British Museum's *Catalogue of Political and Personal Satires*, vol. 8: 1801-1810 (London: British Museum, 1947: 112-14), which comments that the lecturer is "probably not Garnett but Thomas Young", and also identifies many of the people in the audience.

93   **If I had not had to conquer** Quoted in Gillispie: 611.

93   **amusement of hearing Napoleon** Quoted in Hilts: 252.

94   **experimental demonstration** Young, *Miscellaneous Works*, vol. 1: 179.

7 Let There Be Light Waves

95   **The theory of light and colors** Letter to Dalzel (29 Mar. 1802) in Dalzel: 212.

95   **I have ... been accused** Young, *Miscellaneous Works*, vol. 1: 201.

96   **absolutely stationary** Kaku: 11.

96   **The assumption that space** "The fundaments of theoretical physics" in Einstein, *Ideas and Opinions*: 325-26.

97   **sounds are propagated** Newton: 363.

97   **To me the fundamental** Quoted in Wood: 152-53.

99   **If a stone be thrown** Newton: 347-48.

101   **The returns of the disposition** Ibid: 281.

102   **Are not the rays of light** Ibid: 339.

102   **the further I have proceeded** Young, *Miscellaneous Works*, vol. 1: 64.

102   **incomparable writings** Ibid: 78.

103   **one or two difficulties** Ibid: 79.

103   **How happens it that** Ibid: 79.

103   **a still more insuperable difficulty** Ibid: 79.

103   **The phenomena of [these] colors** Ibid: 81.

103   **The greatest difficulty** Ibid: 82.

103   **undoubtedly they cross** Ibid: 83. Young discusses beats in *Natural Philosophy*, vol. 1: 390-91.

104   **What, then, is the analogous** Letter to Gurney (25 June 1830), published as a pamphlet in London in 1830, presumably by Gurney. A copy is in the British Library (shelf mark 8705.a.15).

105   **while reflecting on the beautiful** Young, *Miscellaneous Works*, vol. 1: 202.

105   **Light is probably the undulation** Ibid: 132-33.

105   **Although the invention** Ibid: 140.

106   **A luminiferous ether...not of greater elasticity** Ibid: 142-47.

106   **fundamental** Ibid: 148.

106   **ether distribution hypothesis** Cantor, "The changing role of Young's ether": 44.

106   **When two undulations** Young, *Miscellaneous Works*, vol. 1: 157.

107   **striated surfaces** Ibid: 158.

107   **Suppose a number of** Ibid: 202-03.

108   **Two equal series of waves** Plate 20 in Young, *Natural Philosophy*, vol. 2: 777. Young may have been influenced in developing his theory of interference of light by the interference of the tides at the port of Batsha in the East Indies, judging from his comment in *Natural Philosophy*, vol. 2: 605-06.

111   **In making some experiments** Young, *Miscellaneous Works*, vol. 1: 179.

111   **I made a small hole** Ibid: 179-80.

112   **The foundations of the wave theory** Wood: 168.

8 'Natural Philosophy and the Mechanical Arts'

113 **[The] phenomena of nature** Young, *Natural Philosophy*, vol. 1: 10.

113 **affectionate constancy** Peacock: viii.

114 **It was a marriage of** Ibid: 212.

114 **was happy, though without the comforts** Quoted in Hilts: 253.

114 **Mrs Young has emerged** Letter to Gurney (no date given but must be late Jan. 1811) in Peacock: 221.

114 **Misjudged** Oldham: 56.

115 **As this paper contains nothing** *Edinburgh Review*, vol. 1, (Jan. 1803): 450-52.

116 **In our second number** *Edinburgh Review*, vol. 5, (Oct. 1804): 97.

116 **Conscious of [his] inability** Young, *Miscellaneous Works*, vol. 1: 210-11.

117 **[T]he writer confesses** Ibid: 206-07.

117 **[T]here are two general methods** Ibid: 204.

118 **which I cannot help flattering** Letter to Blagden (15 Sept. 1795) in Cantor, "Henry Brougham and the Scottish methodological tradition": 86.

119 **With this work my pursuit** Young, *Miscellaneous Works*, vol. 1: 215.

119 **it would be difficult to refer** Peacock: 174.

119 **for nearly twenty years** Note by editor (Peacock) in Young, *Miscellaneous Works*, vol. 1: 192.

119 **It is doubtless true** Rayleigh, *Scientific Papers*, vol. 3: 239.

119 **ridicule...his bargain.** Quoted in Wood: 174. Scott's reply is quoted on the following page.

120 **resident physician** Quoted in Wood: 71.

120 **the labor of arranging** Young, *Natural Philosophy*, vol. 1: vi.

121 **a mine to which every one** Gurney: 24-25.

121 **Whether we regard the depth** Kelland: iii.

121 **the greatest and most original** Larmor: 276.

121 **No such authoritative catalog** Ibid: 278.

121 **Reprinting them renders** Young, *Natural Philosophy*, vol. 1: vii.

121 **The weakness of Young's verbal** Ibid: xiii.

122 **[T]he simplest case appears** Young, *Natural Philosophy*, vol. 1: 464-65.

123 **The manner in which two** Young, *Natural Philosophy*, vol. 2: 787.

124 **Young did experiment with** Kipnis: 124. See "The two-slit experiment" in Kipnis: 118-24 for a full discussion of Worrall's argument.

125 **[S]ince the height, to which a body** Young, *Natural Philosophy*, vol. 1: 44.

125 **The term energy** Ibid: 78.

126 **passive strength** Ibid: vii.

126 **[W]e may express the elasticity** Ibid: 137.

126 **This definition is a model** Wood: 131.

126 **Young was the first to introduce** Hetnarski and Ignaczak: 6.

127 **Tearing a liquid column in half** Strutt: 236.

127 **ultimate atoms** Letter to Dominique Arago (12 Jan. 1817) in Oldham: 140.

128 **If heat is not a substance** Young, *Natural Philosophy*, vol. 2: 654.

128 **Important and difficult steps** Peacock: 416-17.

9 Dr Thomas Young, M.D., F.R.C.P.

131 **There is no study more** Young, *An Introduction to Medical Literature*: 2.

131 **in consultations** Letter to Gurney (no date given) in Peacock: 216-17. Oldham: 50 gives the date as 1806.

132 **I purpose seriously** Letter to Gurney (1807) in Peacock: 217-18. Oldham: 50 gives the date as Jan. 1807.

132 **the measurement of minute particles** Young, *Miscellaneous Works*, vol. 1: 343-58. This is a reprint of the section in Young's *An Introduction to Medical Literature*. Wood describes the eriometer on pages 82-86, and illustrates it.

133 **Certainly, science was** Porter, *The Greatest Benefit to Mankind*: 255.

133   **The public regarded him** Arago: 233.

133   **Medical men of all sorts** Porter, *Quacks*: 39.

133   **When any sick** Quoted in Wood: 69.

134   **[I]t is probable that without** Young, *Consumptive Diseases*: 51-52.

134   **the respective advantages** Ibid: 55.

135   **In my own case** Ibid: 62.

135   **With such a combination** Ibid: 63-64.

136   **about half a pint of milk** Ibid: 248-49.

136   **full of accounts** Ibid: 311-12.

136   **The compilation of Bonetus** Ibid: 183-84.

137   **[A]ny of the three candidates** Letter to Gurney (no date given but must be late Jan. 1811) in Peacock: 221.

138   **after a very arduous contest** Quoted in Hilts: 252.

138   **the death of Dr Young** Quoted in Wood: 74.

138   **Dr Young was a great** Peacock: 222.

139   **Medical men, my mood** Letter to Emily Earle (7 Dec. 1815) in Peacock: 250.

139   **It is the peculiar misfortune** Peacock: 216.

139   **The real fact is that** Ibid: 213-14.

140   **Nothing can be more unjust** Brodie: 91-92.

141   **On the functions of the heart** Young, *Miscellaneous Works*, vol. 1: 511-34.

142   **[Young] was not a popular physician** Pettigrew: 9.

10 Reading the Rosetta Stone

143   **You tell me that** Letter to Gurney (no date given) in Peacock: 261. Wood: 210 gives the date as Aug. 1814.

143   **an attempt to unveil** Young, *Some Recent Discoveries in Hieroglyphical Literature*: ix.

143 **first penetrated the obscurity** Quoted in Peacock: 486.

144 **not built up from syllables** Quoted in Boas: 101.

144 **puerile** Young, *Some Recent Discoveries in Hieroglyphical Literature*: 3.

145 **[W]hen they wish to indicate** Boas: 63.

145 **What they mean by a vulture** Ibid: 49-50.

146 **sometimes called the last** *Encyclopaedia Britannica*, 15th edn: entry for "Kircher, Athanasius".

146 **the last man who knew** See the subtitle of the book by Findlen.

146 **The protection of Osiris** Quoted in Pope: 31-32.

146 **according to his interpretation** Young, *Some Recent Discoveries in Hieroglyphical Literature*: 2.

147 **[T]he peculiar nature** Ibid: 7.

148 **notae phoneticae** Pope: 58.

148 **it halted of itself** Quoted in Claiborne: 24.

149 **is the most popular** Parkinson, *The Rosetta Stone*: 47.

149 **Unfortunately, the stone's iconic** Parkinson, *Cracking Codes*: 25.

151 **This decree shall be** Quoted in Andrews: 28.

153 **[They] proceeded upon** Young, *Some Recent Discoveries in Hieroglyphical Literature*: 9.

154 **in which he asserted** Ibid: xiv-xv.

155 **those who have not been in the habit** Young, *Miscellaneous Works*, vol. 3: 612.

156 **It is impossible to form** Peacock: 281. The five volumes of Young's MSS in the British Library have the shelf mark Add. 27281-27285.

156 **striking resemblance** Letter to Sylvestre de Sacy (3 Aug. 1815) in Young, *Miscellaneous Works*, vol. 3: 54.

156 **I discovered, at length** Young, *Some Recent Discoveries in Hieroglyphical Literature*: 15-16.

157   **I am not surprised that** Letter to de Sacy (3 Aug. 1815) in Young, *Miscellaneous Works*, vol. 3: 53.

157   **imitations of the hieroglyphics** Ibid: 54.

158   **it seemed natural to suppose** Young, *Miscellaneous Works*, vol. 3: 133.

158   **If I might venture to advise you** Letter to Young (20 July 1815) in Peacock: 266-67. The French original appears in Young, *Miscellaneous Works*, vol. 3: 51.

159   **Since Champollion was obsessed** Ray, "The name of the first: Thomas Young and the decipherment of Egyptian writing" (unpublished lecture).

160   **The square block** Young, *Miscellaneous Works*, vol. 3: 156-57.

161   **Egypt** Young's article for the *Encyclopaedia Britannica* appears in Young, *Miscellaneous Works*, vol. 3: 86-197; the vocabulary is at the end on fold-out sheets.

161   **mixed up with** Quoted in Gardiner: 14.

163   **I have the satisfaction** Belzoni: 205-06.

11 Waves of Enlightenment

165   **I dare say poor Fresnel** Letter to Emily Earle (Nov. 1827) in Peacock: 397.

166   **It seems as though, as it passes** Quoted in Andriesse: 273.

167   **The most singular** Young, *Natural Philosophy*, vol. 1: 445.

168   **Have not the rays** Newton: 358.

169   **Are not all hypotheses** Ibid: 361.

169   **It would be like trying** Zajonc: 117.

170   **In the intermediate period** Peacock: 369-70.

171   **This statement appears to us** Quoted in Wood: 181.

171   **With respect to my own** Letter to Brewster (13 Sept. 1815) in Young, *Miscellaneous Works*, vol. 1: 361.

172   **But notwithstanding all** Young, *Miscellaneous Works*, vol. 1: 279-80.

172   **structure of the elementary atoms** Ibid: 228.

173 **blinded** Wood: 186.

173 **It is certainly easier** Young, *Miscellaneous Works*, vol. 1: 332.

173 **I have … been reflecting** Letter to Arago (12 Jan. 1817) in Young, *Miscellaneous Works*, vol. 1: 383.

176 **This hypothesis of Mr Fresnel** Young, *Miscellaneous Works*, vol. 1: 415. Fresnel's work is thoroughly discussed by Kipnis and by Buchwald.

176 **if anything could console me** Letter to Young (24 May 1816) in Young, *Miscellaneous Works*, vol. 1: 378. This translation appears in Wood: 189.

177 **In the year 1816** Arago: 239-40.

178 **The unpursued speculations** Quoted in Whewell: 349-50.

178 **[W]e must not, in our regard** Quoted in Wood: 204.

178 **has divided the prize** Letter to Emily Earle (Nov. 1827) in Peacock: 397.

12 Walking Encyclopedia

179 **The longer a person has lived** Letter to Gurney (June? 1809) in Peacock: 220.

180 **under other circumstances** Letter to Napier (9 Aug. 1814) in Peacock: 252.

180 **I have long been intending** Letter to Emily Earle (22 Nov. 1814) in Peacock: 246.

181 **I could not at present allow** Letter to Napier (12 Feb. 1816) in Peacock: 253.

181 **as I got nothing** Letter to Gurney (3 Jan. 1821) in Wood: 90.

182 **was continually trying to foist** Wood: 258. This chapter was written by Oldham.

182 ***Baths* I cannot refuse** Letter to Napier (Feb. 1816) in Wood: 258.

182 **For the last ten years** Letter to Napier (no date given) in Wood: 258.

182 **There was a time in my life** Letter to Napier (30 Jan. 1821) in Wood: 258-59.

182 **No subject comes amiss** Letter to Young (27 Apr. 1823) in Wood: 263.

182 **I shall not even object** Letter to Napier (30 Apr. 1823) in Wood: 263.

182 **the only one I did** *con amore* Letter to Napier (no date given) in Wood: 270.

182 **Lagrange will be** Letter to Napier (no date given) in Wood: 270. See Young's letter to Gurney, 18 Dec. 1820, mentioning his article on Lagrange, in Wood: 318.

183 **The biographical articles** Letter to Gurney (no date given) in Wood: 270-71.

183 **I have little patience** Einstein, *The New Quotable Einstein*: 247.

183 **to whose profound and accurate** Quoted in Wood: 271.

184 **Of language in general** Young, *Miscellaneous Works*, vol. 3: 480-81.

185 **every one of them has** Ibid: 513.

186 **what is the historical reality** Renfrew: 11.

186 **although I well knew** Letter to Peacock (no date given) in Young, *Miscellaneous Works*, vol. 2: 262.

186 **In the theory of tides** Rayleigh, "Commemoration lecture": 206.

187 **he has hinted at** Letter to Peacock (no date given) in Young, *Miscellaneous Works*, vol. 2: 262.

187 **deduced some important** Wood: 260. This chapter was written by Oldham.

187 **I was surprised to find how far** Quoted in Wood: 260.

187 **the only reasonable doubt relates** Young, *Miscellaneous Works*, vol. 2: 246.

13 In the Public Interest

189 **The cultivation of abstract science** Quoted in Wood: 312.

189 **The principal timbers** Peacock: 345.

190 **I have been excommunicated** Letter to Young (no date given) in Peacock: 347.

190 **Many of the old captains** Peacock: 347.

190 **I ought perhaps to have returned** Letter to Barrow (22 Nov. 1811) in Young, *Miscellaneous Works*, vol. 1: 536.

191   **It appears, therefore, to be** Young, *Miscellaneous Works*, vol. 1: 552-53.

191   **It is by no means impossible** Ibid: 561.

191   **Dr Young was not easily seduced** Peacock: 348.

191   **Though science is much respected** Quoted in Peacock: 349. The official is not named by Peacock.

191   **He cannot, we think** Quoted in Peacock: 349. Barrow's review was published in 1815.

192   **It was a problem of a very high order** Peacock: 356.

193   **It seems right to state** Gurney: 33.

193   **The length of a simple pendulum** Wood: 290. This chapter was written by Oldham.

194   **George Graham** Young, *Miscellaneous Works*, vol. 2: 430.

194   **with great ingenuity** Ibid: 433.

194   **There could be no doubt** Quoted in Hilts: 253.

196   **It is not easy to define** Peacock: 359.

196   **[I]t is difficult for the warmest admirers** Ibid: 365.

196   **unscrupulous methods** Edmund Dews, "Thomas Young as a civil servant", in Wood: 345.

196   **The price finally paid** Ibid: 345.

197   **headed a reform campaign** Schaffer: 275.

197   **If every practical astronomer were** Letter to Airy (no date given) in Peacock: 362-63.

197   **an act of barbarism** Peacock: 364.

198   **An orator** Arago: 235-36.

198   **highly conducive** Letter to John Barrow (22 July 1820) in Wood: 306.

199   **And here is the *polar expedition*** Letter to Gurney (6 Nov. 1820) in Wood: 307.

14 Grand Tour

201 **[Our expedition] seems like** Letter to Gurney (8 Sept. 1821) in Wood: 323. The date is given in Wood: 321.

201 **in the summer of 1821** Quoted in Hilts: 253.

202 **block of granite** Young, *Some Recent Discoveries in Hieroglyphical Literature*: 28.

202 **by far the most illustrious** Peacock: 453.

203 **We were delighted** Letter to Gurney (8 July 1821) in Peacock: 454. The date is given in Wood: 321.

203 **jewelry** Quoted in Adkins: 238.

204 **the excavations exploited** Gardiner: 17.

204 **Pisa amply repaid** Letter to Gurney (8 Sept. 1821) in Wood: 322-23. The date is given in Wood: 321.

205 **Whatever may be Mr Drovetti's decision** Letter to MM. Mompurgo (5 Sept. 1821) in Young, *Some Recent Discoveries in Hieroglyphical Literature*: 36.

205 **Mr Drovetti's cupidity** Young, *Some Recent Discoveries in Hieroglyphical Literature*: 37-38.

206 **On the whole our expedition** Letter to Gurney (8 Sept. 1821) in Wood: 323. The date is given in Wood: 321.

206 **delighted us extremely** Letter to Gurney (8 Sept. 1821) in Peacock: 456. The date is given in Wood: 321.

206 **We fell into a good deal of society** Ibid: 456.

206 **Of the science and literature of this country** Ibid: 458-59.

206 **hasten home** Letter to Gurney (8 Sept. 1821) in Wood: 323. The date is given in Wood: 321.

207 **boyhood** Ibid: 323.

15 Dueling with Champollion

209   **Mr Champollion, junior** Letter to Hamilton (29 Sept. 1822) in Young, *Miscellaneous Works*, vol. 3: 220.

210   **So poor Dr Young is incorrigible?** Letter to Champollion-Figeac (25 Mar. 1829) in Champollion: 184.

210   **the suspicion may easily arise** Ray, "The name of the first: Thomas Young and the decipherment of Egyptian writing" (unpublished lecture).

211   **could not bear** Gurney: 46.

211   **for employing some poor** Letter to Gurney (Oct. 1817) in Peacock: 451.

212   **I almost entirely undressed** Letter to Champollion-Figeac (1 Jan. 1829) in Champollion: 140.

212   **while maintaining civil relations** Adkins: 190.

213   **simple modification** Quoted in Solé and Valbelle: 76.

214   **as I had not leisure** Young, *Some Recent Discoveries in Hieroglyphical Literature*: 49. Bankes's contribution to the decipherment is discussed in Usick: 77-79.

217   **Fresnel, a young mathematician** Letter to Gurney (Sept. 1822) in Peacock: 321-22 and Wood: 231. Both Peacock and Wood quote the letter but in somewhat different ways.

217   **I did certainly expect** Young, *Some Recent Discoveries in Hieroglyphical Literature*: 43.

218   **The hieroglyphical text** Quoted in Young, *Some Recent Discoveries in Hieroglyphical Literature*: 44-45.

218   **This course of investigation** Young, *Some Recent Discoveries in Hieroglyphical Literature*: 45-46.

219   **I shall never consent** Letter to Young (23 Mar. 1823) in Young, *Miscellaneous Works*, vol. 3: 256. The translation is from Wood: 237.

219   **Nothing can exceed** Note by editor (Leitch) in Young, *Miscellaneous Works*, vol. 3: 255.

220 **[that] the further [Champollion] advances** Young, *Some Recent Discoveries in Hieroglyphical Literature*: 53-54.

221 **I recognize that he was the first** Quoted in Parkinson, *Cracking Codes*: 40.

222 **He found that it is easier** Quoted in Hilts: 253.

16 A Universal Man

223 **He might for example** Quoted in Hilts: 253.

224 **[H]is own idea was** Ibid: 253-54.

225 **adequate to his utmost wishes** Ibid: 253.

225 **A rare example of conscientiousness** Peacock: 404.

225 **It is obvious that those counties** Young, *Consumptive Diseases*: 105-06.

226 **with greater prudence than equity** Peacock: 413.

226 **I sincerely hope that** Young, *Miscellaneous Works*, vol. 2: 377-78.

227 **The public have lately** Quoted in Young, *Miscellaneous Works*, vol. 2: 380.

227 **where the agitation produced** Young, *Natural Philosophy*, vol. 2: 717.

227 **the shape assumed** Craik: 232.

228 **Until now, mathematicians** Quoted in Young, *Miscellaneous Works*, vol. 2: 79.

228 **M. Laplace's hypothesis** Young, *Miscellaneous Works*, vol. 2: 81.

228 **from a single tangent** Young's paper, "Determination of the figure of the earth from a single tangent" appears in Peacock: 511-14.

228 **it is my pride and pleasure** Letter to Gurney (17 Dec. 1828) in Wood: 328.

229 **I could not, therefore, but conclude** Young, *Some Recent Discoveries in Hieroglyphical Literature*: 58.

230 **You write that from time to time** Letter to Young (28 May 1827) in Young, *Miscellaneous Works*, vol. 3: 423-24.

230 **has shown me far more attention** Letter to Gurney (no date given) in Wood: 247. See also Peacock: 341.

231 **thirty years ago** "Advertisement" in Young, *Rudiments of an Egyptian Dictionary*: vii.

231 **Young was the first person** Ray, "The name of the first: Thomas Young and the decipherment of Egyptian writing" (unpublished lecture).

231 **Knowledge exists in two forms** "Message in honor of Morris Raphael Cohen" in Einstein, *Ideas and Opinions*: 80.

232 **I have derived more pleasure** Letter to Gurney (28 Dec. 1820) in Wood: 320.

232 **I have been dashing through** Letter to Emily Earle (no date given) in Peacock: 215-16.

232 **one of the very few dinners** Letter to Emily Earle (22 Nov. 1814) in Peacock: 247-49.

233 **I find there has been pretty general** Letter to Emily Earle (no date given) in Wood: 326.

233 **I told him that he had not quite enough** Quoted in Peacock: 474.

234 **he led the life of a philosopher** Gurney: 37-38.

234 **the pursuit of such fame** Quoted in Gurney: 38.

234 **Living in a period** Cantor, "Thomas Young": 949.

234 **As far as the qualities of the mind** Quoted in Hilts: 254.

235 **that nothing should go forth** Gurney: 42.

235 **that it was a work which** Ibid: 42.

235 **The parietes of the heart** "Examination of the body of the late Thomas Young M.D. For. Sec. R.S. May 11 1829", St George's Hospital Library, London. The MS is a short, two-page note in the handwriting of Brodie (courtesy of the librarian Nallini Thevakarrunai).

236 **unwearied and incessant labor** Gurney: 44.

236 **Ossification of the aorta** Personal communication from David Sprigings, Sept. 2005.

236 **The death of Young in his own country** Arago: 236.

237 **distinguished physician** *Lancet*, vol. 2, (23 May 1829): 255.

237 **The multiplied objects** Quoted in Peacock: 482-83.

238 **how inadequate and limited a view** Letter to Gurney (25 June 1830), published as a pamphlet in London in 1830, presumably by Gurney. A copy is in the British Library (shelf mark 8705.a.15).

Bibliography

Books

Adkins, Lesley and Roy, *The Keys of Egypt: The Race to Read the Hieroglyphs*, London: HarperCollins, 2000

Anderson, Verily, *Friends and Relations: Three Centuries of Quaker Families*, London: Hodder and Stoughton, 1980

Andrews, Carol, *The Rosetta Stone*, London: British Museum Publications, 1981

Andriesse, C. D., *Huygens: The Man Behind the Principle*, Cambridge (UK): Cambridge University Press, 2005

Belzoni, Giovanni, *Belzoni's Travels: A Narrative of the Operations and Recent Discoveries in Egypt and Nubia*, Alberto Siliotti ed., London: British Museum Press, 2001

Bence Jones, Henry, *The Royal Institution: Its Founders, and Its First Professors*, London: Longmans, Green, 1871

Boas, George, trans., *The Hieroglyphics of Horapollo*, pbk edn, Princeton: Princeton University Press, 1993

Brodie, Benjamin, *The Works of Sir Benjamin Collins Brodie*, collected and arranged by Charles Hawkins, 3 vols, London: Longmans, 1865

Buchwald, Jed Z., *The Rise of the Wave Theory of Light: Optical Theory and Experiment in the Early Nineteenth Century*, Chicago: University of Chicago Press, 1989

Champollion, Jean-François, *Egyptian Diaries: How One Man Solved the Mysteries of the Nile*, London: Gibson Square Books, 2001

Chapman, Allan, *Mary Somerville and the World of Science*, Bristol (UK): Canopus, 2004

Claiborne, Robert, *The Birth of Writing*, New York: Time-Life Books, 1974

Dalzel, Andrew, *History of the University of Edinburgh, from its Foundation: with a Memoir of the Author*, Edinburgh: Edmonston and Douglas, 1862 (memoir of Dalzel by C. Innes)

Duleep Singh, Frederick, *Portraits in Norfolk Houses*, Edmund Farrer ed., vol. 1, Norwich (UK): Jarrold and Sons, 1928—gives details of the portrait of Thomas Young by Lawrence

Einstein, Albert:

> *Ideas and Opinions*, Carl Seelig ed., pbk edn, New York: Three Rivers Press, 1982

> *The New Quotable Einstein*, Alice Calaprice ed., Princeton: Princeton University Press, 2005

El-Daly, Okasha, *Egyptology: The Missing Millennium: Ancient Egypt in Medieval Arabic Writings*, London: UCL Press, 2005

Feynman, Richard, Robert Leighton and Matthew Sands, *The Feynman Lectures on Physics*, vol. 1, Reading (US): Addison-Wesley, 1965

Findlen, Paula, ed., *Athanasius Kircher: The Last Man Who Knew Everything*, London: Routledge, 2004

Gardiner, Alan, *The Egyptians: An Introduction*, London: Folio Society, 1999 (originally published in 1961)

Gillispie, Charles Coulston, *Science and Polity in France: The Revolutionary and Napoleonic Years*, Princeton: Princeton University Press, 2004

Godfrey, Richard, *James Gillray: The Art of Caricature*, London: Tate Publishing, 2001

Hahn, Roger, *Pierre-Simon Laplace 1749-1827: A Determined Scientist*, Cambridge (US): Harvard University Press, 2005

Helmholtz, Hermann, *Popular Lectures on Scientific Subjects*, London: Longmans, Green, 1873

Hetnarski, Richard B. and Józef Ignaczak, *Mathematical Theory of Elasticity*, London: Taylor and Francis, 2004

Hubel, David, *Eye, Brain, and Vision*, pbk edn, New York: Freeman (Scientific American Library), 1995

James, Frank A. J. L., ed., *'The Common Purposes of Life': Science and Society at the Royal Institution of Great Britain*, Aldershot (UK): Ashgate, 2002

Kaku, Michio, *Einstein's Cosmos: How Albert Einstein's Vision Transformed Our Understanding of Space and Time*, London: Weidenfeld and Nicolson, 2004

Kipnis, Nahum, *History of the Principle of Interference of Light*, Basle (Switzerland): Birkhäuser, 1991

Kline, Daniel Louis, *Thomas Young, Forgotten Genius: An Annotated Narrative Biography*, Cincinnati (US): Vidan Press, c. 1993

Moore, Walter, *Schrödinger: Life and Thought*, Cambridge (UK): Cambridge University Press, 1989

Moore, Wendy, *The Knife Man: The Extraordinary Life and Times of John Hunter, Father of Modern Surgery*, London: Bantam, 2005

Murray, Alexander, ed., *Sir William Jones 1746-94: A Commemoration*, Oxford: Oxford University Press, 1998

Newton, Isaac, *Opticks*, London: G. Bell and Sons, 4th edn, 1931 (with a foreword by Albert Einstein)

Oldham, Frank, *Thomas Young, F.R.S.: Philosopher and Physician*, London: Edward Arnold, 1933

Park, David, *The Fire in the Eye: A Historical Essay on the Nature and Meaning of Light*, Princeton: Princeton University Press, 1997

Parkinson, Richard:

    Cracking Codes: The Rosetta Stone and Decipherment, London: British Museum Press, 1999

    The Rosetta Stone, London: British Museum Press, 2005

Peacock, George, *Life of Thomas Young, M.D., F.R.S.*, London: John Murray, 1855

Pope, Maurice, *The Story of Decipherment: From Egyptian Hieroglyphs to Maya Script*, rev. edn, London: Thames and Hudson, 1999

Porter, Roy:

   London: A Social History, pbk edn, London: Penguin, 1996

   The Greatest Benefit to Mankind: A Medical History of Humanity from Antiquity to the Present, London: HarperCollins, 1997

   Quacks: Fakers and Charlatans in Medicine, pbk edn, Stroud (UK): Tempus, 2001

Rayleigh, Lord (John William Strutt), *Scientific Papers*, 6 vols, Cambridge: Cambridge University Press, 1899-1920

Renfrew, Colin, *Archaeology and Language: The Puzzle of Indo-European Origins*, London: Cape, 1987

Robinson, Andrew:

   The Story of Writing: Alphabets, Hieroglyphs and Pictograms, London: Thames and Hudson, 1995

   Lost Languages: The Enigma of the World's Undeciphered Scripts, New York: McGraw-Hill, 2002

   The Man Who Deciphered Linear B: The Story of Michael Ventris, London: Thames and Hudson, 2002

Sobel, Dava, *Longitude: The True Story of a Lone Genius Who Solved the Greatest Scientific Problem of His Time*, London: Fourth Estate, 1996

Solé, Robert and Dominique Valbelle, *The Rosetta Stone: The Story of the Decoding of Hieroglyphics*, London: Profile, 2001

Somerville, Martha, *Personal Recollections from Early Life to Old Age of Mary Somerville*, London: John Murray, 1873

Strutt, Robert John, *Life of John William Strutt, Third Baron Rayleigh, O.M., F.R.S.*, Madison: University of Wisconsin Press, 1968

Wade, Nicholas J.:

   A Natural History of Vision, Cambridge (MA): MIT Press, 1998

   Perception and Illusion: Historical Perspectives, New York: Springer, 2005

Westfall, Richard S., *The Life of Isaac Newton*, Cambridge (UK): Cambridge University Press, 1993

Whewell, William, *History of the Inductive Sciences, From the Earliest to the Present Time*, vol. 2, 3rd edn, London: Frank Cass, 1967 (new impression, with index, of the 1857 edn)

Wood, Alex, *Thomas Young: Natural Philosopher 1773-1829*, Cambridge (UK): Cambridge University Press, 1954 (completed by Frank Oldham)

Young, Thomas:

> *An Introduction to Medical Literature, Including a System of Practical Nosology*, London: W. Phillips, 1813

> *A Practical and Historical Treatise on Consumptive Diseases*, London: Thomas Underwood and John Callow, 1815

> *An Account of Some Recent Discoveries in Hieroglyphical Literature, and Egyptian Antiquities*, London: John Murray, 1823

> *Rudiments of an Egyptian Dictionary in the Ancient Enchorial Character*, London: J. and A. Arch, 1831

> *A Course of Lectures on Natural Philosophy and the Mechanical Arts*, 4 vols, Bristol (UK): Thoemmes, 2002 (with an introduction by Nicholas J. Wade)—facsimile reprint of the original two-volume edition published in London in 1807 by Joseph Johnson

> *Miscellaneous Works of the Late Thomas Young, M.D., F.R.S.*, George Peacock ed. (vols 1 and 2) and John Leitch ed. (vol. 3), Bristol (UK): Thoemmes, 2003—facsimile reprint of the original edition published in London in 1855 by John Murray

Zajonc, Arthur, *Catching the Light: The Entwined History of Light and Mind*, London: Bantam, 1993

Articles

Arago, Dominique-François, "Biographical memoir of Dr Thomas Young", *Edinburgh New Philosophical Journal*, vol. 20, (Apr. 1836): 213-40

Asimov, Isaac, "Thomas Young", in *Asimov's Biographical Encyclopedia of Science and Technology*, rev. edn, Garden City (US): Doubleday, 1972: 241-42

Boycott, Brian B., "John Zachary Young", *Biographical Memoirs of Royal Society Fellows*, London: Royal Society, 1998: 487-509

Brooke, John and Geoffrey Cantor, "'A taste for philosophical pursuits'—Quakers in the Royal Society of London", in Brooke and Cantor, *Reconstructing Nature: The Engagement of Science and Religion*, New York: Oxford University Press, 2000: 282-313

[Brougham, Henry], review of three lectures by Thomas Young, *Edinburgh Review*, vol. 1, (Jan. 1803): 450-60, and vol. 5, (Oct. 1804): 97-103

Cantor, Geoffrey:

"The changing role of Young's ether", *British Journal for the History of Science*, vol. 5, (1970): 44-62

"Thomas Young's lectures at the Royal Institution", *Notes and Records of the Royal Society of London*, vol. 25, (1970): 87-112

"Henry Brougham and the Scottish methodological tradition", *Studies in the History and Philosophy of Science*, vol. 2, (1971): 69-89

"The reception of the wave theory of light in Britain: a case study illustrating the role of methodology in scientific debate", *Historical Studies in the Physical Sciences*, vol. 6, (1975): 109-32

"Thomas Young", in *Oxford Dictionary of National Biography*, 60, Oxford: Oxford University Press, 2004-05: 945-49

Craik, Alex D. D., "James Ivory, F.R.S., mathematician: 'the most unlucky person that ever existed'", *Notes and Records of the Royal Society of London*, vol. 54, (2000): 223-47

Fonda, Gerald, "Bicentenary of the birth of Thomas Young, M.D., F.R.S.", *British Journal of Ophthalmology*, vol. 57, (1973): 803-08

[Gurney, Hudson], "Memoir [of Young]", in *Thomas Young, Rudiments of an Egyptian Dictionary*: 5-47

Hilts, Victor L., "Thomas Young's 'Autobiographical sketch'", *Proceedings of the American Philosophical Society*, vol. 122, (Aug. 1978): 248-60

Kelland, Philip, "Preface by the editor", in *Thomas Young, A Course of Lectures on Natural Philosophy and the Mechanical Arts*, 2nd edn, London: Taylor and Walton, 1845: iii-v

Lancet, The, "Dr Young", vol. 2, (23 May 1829): 255

Larmor, Joseph, "Thomas Young", *Nature*, vol. 133, (24 Feb. 1934): 276-79

Lees, Charles Herbert, "Thomas Young", in *Dictionary of National Biography*, Oxford: Oxford University Press, 1997: 1308-14 (entry originally published in 1900)

Levene, John, "Sir George Biddell Airy, F.R.S. (1801-1892) and the discovery and correction of astigmatism", *Notes and Records of the Royal Society of London*, vol. 21, (1966): 180-99

Morse, Edgar W., "Thomas Young", in Charles Coulston Gillispie ed., *Dictionary of Scientific Biography*, vol. 14, New York: Scribner, 1976: 562-72

Oldham, Frank, "Thomas Young", *British Medical Journal*, 19 Oct. 1974: 150-52

Pettigrew, Thomas Joseph, "Thomas Young, M.D., F.R.S.", in Pettigrew, *Biographical Memoirs of the Most Celebrated Physicians, Surgeons*, London: Whittaker, 1839: 1-24

Rayleigh, Lord, "Commemoration lecture", *Journal of the Royal Institution*, 1899-1902: 204-06

Robinson, Herbert Spencer, "Thomas Young: a chronology and a bibliography with estimates of his work and character", *Medical Life*, vol. 36, (1929): 527-40

Rowell, H. S., "Thomas Young and Göttingen", *Nature*, vol. 88, (15 Feb. 1912): 516

Rubinowicz, A., "Thomas Young and the theory of diffraction", *Nature*, vol. 180, (27 July 1957): 160-62

Schaffer, Simon, "Paper and brass: the Lucasian professorship 1820-39", in *From Newton to Hawking: A History of Cambridge University's Lucasian Professors of Mathematics*, Kevin C. Knox and Richard Noakes eds, Cambridge (UK): Cambridge University Press, 2003: 241-93

Thomas, John Meurig, "Sir Benjamin Thompson, Count Rumford and the Royal Institution", *Notes and Records of the Royal Society of London*, vol. 53, (1999): 11-25

Index

The main references are in **bold**. TY stands for Thomas Young. All places and all institutions are located in the United Kingdom, unless otherwise indicated.